Praise for *We're All Climate Hypocrites Now*

A useful—and sprightly!—effort to get at the cho~~~ ~ dividual and systemic action on the great~ faced. I found it a helpful spur to creative th I bet you will as well. Read it, and then get the politics and economics that are driving u~ hell, then a place with a similar temperature.

— Bill McKibben, author, *The End of Nature*

We're All Climate Hypocrites Now is part eco-therapy, part climate strategy, and a fantastic antidote to the overwhelm that comes along with living in a global ecological crisis. Say goodbye to those little voices in your head (or those loud voices on Facebook) calling you a hypocrite because you don't bike to work, aren't vegan, fly to a protest, and still haven't taken out that loan for those rooftop solar panels. This book is a fresh and informative unpacking of why we must abandon the notion that individual eco-perfection is possible—or even impactful—in the absence of system-wide change. It's an inspiring call to let go of the "either or" mentality, to fully embrace the "both and," and to remember to go easy on ourselves and each other as we lean in even further into this painful, chaotic yet exciting time of (r)evolution.

— Danna Smith, executive director, Dogwood Alliance

Sami Grover's wise book charts a middle way to win transformational change. He challenges us to embrace our climate hypocrisy as a goal to uproot the structures that are killing the planet without losing sight of the strategic individual actions we can take right now. We can't curate our way out of the climate crisis as consumers—we must replace the system that makes us climate hypocrites. We climate hypocrites have agency, in varying degrees, to take actions that multiplied by the millions will help to win the big changes we need to survive. With our eyes on the stars and our feet on the ground, we can meet ourselves where we are without guilt and act for a more equitable, just, and sustainable world. Let this book show you how.

— Bill Corcoran, Sierra Club's Beyond Coal campaign

What a great book. Grover pushes well beyond BTUs and solar installs to confront shame, duplicity, and the multi-ality of being human. It's daring, bold, and wonderfully provocative. One moment I'm hoping he buys the new crepe pan, the next I'm staring in the mirror thinking about my wasteful habits. It's a great read with an epic span—from the morality of procreation to a wheelbarrow of horse shit and back again. Loved it.

— Lyle Estill, author, *Small Is Possible*, grandparent, distiller

If you are a climate concerned person who struggles with the nuanced complexity of being "green," Sami's book will help you navigate this contemporary moral maze with intelligent bigger picture thinking plus a rich seam of strategies and initiatives large and small for a healthier planet.

— Maddy Harland, co-founder & editor, *Permaculture Magazine*, author, *Fertile Edges*

Nobody knows more about the business of sustainability than Sami Grover. He brings a welcome dose of wit, clarity, and levity to the green movement.

— Brian Merchant, best-selling author, *The One Device*

On every page of this rip-roaring read I found myself, my partner, my neighbour, my colleagues, my family, and my friends and every holier-than-thou temptation, every emptying out of the compost bin, every person who berated me for traveling for work with refugees. Hypocrisy is in our DNA, and in this book it is both hilariously observed, with all the dry wit of a Brit, and pragmatically harnessed for good. I honestly could not put it down. It's a *tour de force* for hope. And kindness. And love for the world and the future.

— Alison Phipps, UNESCO Chair for Refugee Integration through Languages and the Arts, University of Glasgow

Grover's nuanced take on how to approach life at the end of the world—with thoughtfulness, honesty, and more than a touch of pragmatism—gives even the most jaded amongst us a boost of energy to do our part in extending the health of the planet, for as long as we can. While he turns to science, public commentary, experts, and activists to remind us that individual energies and collective action can produce results, what makes *We're All Climate Hypocrites Now* is Grover's willingness to put himself squarely in the middle of these debates, engaging with his own evolution as a climate activist, and evaluating and reevaluating his own practices. The honesty with which he writes is an invitation, rather than an edict, to join him in making a difference.

— Dr. Kumarini Silva, associate professor, Communication and Cultural Studies, UNC-Chapel Hill, author, *Brown Threat*

Sami Grover's *We're All Climate Hypocrites Now* is an enjoyable, well-timed tonic for often-bitter debates about whether a focus on personal emissions helps or hurts societal efforts to stop the climate crisis. His proposal is to repurpose Big Oil's buck passing "carbon footprint" into a tool to identify your own specific leverage to create wider social change. A thinking person's call to action, best enjoyed with a cold, well-crafted American beer.

— Dr. Dan Rutherford, Aviation Director, International Council on Clean Transportation

As someone who teaches environmental advocacy, I know the paralysis, guilt, and self-blame that sometimes hits us when the odds seem insurmountable, and Grover's book is a great antidote. Grover reminds us that the social, economic, and political roots of climate change are broad and deep, but so are the solutions. With in-depth interviews and entertaining anecdotes, he shows us how climate scientists, activists, and advocates are finding ways to address the myriad problems that are linked to the unfolding climate crisis. Grover's candid and upbeat approach offers a fresh and inspiring take on climate action that will become required reading in my courses.

— Dr. David Monje, assistant professor, co-director, UNC's Program for Cultural Studies

A thought-provoking, insightful, and witty exploration of the familiar dilemmas we face navigating the world of climate solutions, from personal choices to systemic change, and the interplay between the two.

— Andreas Karelas, founder and executive director,
RE-volv, author, *Climate Courage*

Over the years, many have struggled mightily to find a place of balance and comfort in the pursuit of personal accountability in their individual lives, while addressing the ultimate solutions needed to conquer the carbon crisis, which is at the root of climate change. Sami highlights a multitude of actions on a variety of scales that trigger this consternation while appropriately calling-out the systems of greatest causation where the focus belongs. His wit, insightfulness, and informed knowledge is refreshing and on full display. Sami's "judge not, that you be not judged" accounts in the book are humorous, thought provoking, and profound. *We're All Climate Hypocrites Now* is a must read for all climate change warriors.

— Joe Jackson, board vice chair, Dogwood Alliance,
founder, EcoGrounds Management Systems

We're All Climate Hypocrites Now

*How Embracing Our Limitations Can
Unlock the Power of a Movement*

Sami Grover

new society
PUBLISHERS

Cover design by Diane McIntosh.
Cover images: ©iStock

Printed in Canada. First printing September 2021.

Inquiries regarding requests to reprint all or part of *We're All Climate Hypocrites Now* should be addressed to New Society Publishers at the address below. To order directly from the publishers, please call toll-free (North America) 1-800-567-6772, or order online at www.newsociety.com

Any other inquiries can be directed by mail to:

New Society Publishers
P.O. Box 189, Gabriola Island, BC V0R 1X0, Canada
(250) 247-9737

LIBRARY AND ARCHIVES CANADA CATALOGUING IN PUBLICATION

Title: We're All climate hypocrites now : how embracing our limitations can unlock the power of a movement / Sami Grover.

Other titles: We are all climate hypocrites now

Names: Grover, Sami, author.

Description: Includes bibliographical references and index.

Identifiers: Canadiana (print) 20210265116 | Canadiana (ebook) 20210265183 | ISBN 9780865719606

(softcover) | ISBN 9781550927535 (PDF) | ISBN 9781771423496 (EPUB)

Subjects: LCSH: Environmental responsibility. | LCSH: Environmental protection—Citizen participation. | LCSH: Environmentalism.

Classification: LCC GE195.7 .G76 2021 | DDC 363.7—dc23

New Society Publishers' mission is to publish books that contribute in fundamental ways to building an ecologically sustainable and just society, and to do so with the least possible impact on the environment, in a manner that models this vision.

For Lilia, Adeline, and Jenni.
You're why I do what I do.
(Now turn out the bloody lights.)

Contents

Acknowledgments:
An Incomplete Catalog of
Gushing Praise and Profuse Thanks

When I first started writing this book, I set aside some budget for travel. After all, while not exactly great for my carbon footprint, I figured I would need to go meet with people so that I could truly understand the nature of the topic I was delving into.

That was not to be. And neither did it prove to be necessary.

The coronavirus lockdowns forced me to try something different—and I will forever be profoundly grateful and a little surprised at the sheer number and quality of enlightening, entertaining, and inspiring conversations I had with experts, authors, academics, and activists in so many corners of the world, all without getting on a plane or unnecessarily frying the atmosphere. A huge thanks to all of them—and to the many other writers, thinkers, and doers whose work has informed my peculiar ramblings.

Speaking of peculiar ramblings, this book would not have been possible without the 30+ years of do-gooding that preceded it. So a thanks to all the various individuals and groups I've interacted with since my early teens, including but not limited to the Bristol Permaculture crowd (special, big hugs to Sarah Pugh) and Ragman's Lane Farm folks; the good people at Treesponsibility; Clevedon & North Somerset Friends of the Earth; Hull Hempology and People & Planet; Hitchcock's Vegetarian Restaurant; Durham Living Wage Project; Tami and The Abundance Foundation crew; my Redwoods, *Treehugger*, Multilingual/Channel View and Change Creation co-conspirators, and of course the awesome folks at Dogwood Alliance. (I'd also like to express my appreciation to all my old teenage "wombling" mates that helped me pick up our litter after yet another misspent night at Ladye Bay.) Oh, and while we're at it, let's offer up a big fat middle finger to all the powerful forces who continue to stand in the way of progress.

But I digress.

Huge and explicit thanks go to those who made this book happen: to Lyle Estill, for the introductions (it was worth every single ounce of apple brandy); to Ingrid Witvoet, Rob West, Murray Reiss, Sue Custance, and everyone at New Society; to Kumi and David for the thorough review and genuinely transformative edits; to Think Club, past and present, for never quite getting around to reviewing, but offering excellent brain food nonetheless; to Supper Club for the nourishment and conversations; and, of course to various family members who cast their eye over parts of it—Tommi, Sara, Mum!!, Sarah, Kris (sorry for not sending it to you in time LaLa and Wesley!). And while not exactly book related, I'd also like to offer a thanks to my Finnish family for the formative experiences in what must surely be one of the most beautiful parts of the world.

Biggest and most profuse thanks go to Jenni, Lilia, and Adeline—for putting up with my litter picking at the beach while they collect shells; for supporting, encouraging and offering excellent and insightful thoughts on this book ("I'd never buy it, but it's actually not bad!"); and for forever ruining my carbon footprint by moving me to the other side of the pond.

And because I have almost certainly inadvertently missed some folks, I am going to offer a single, blanket thank you to everyone whom—however imperfectly—has even started to engage with what is often a terrifying, depressing, and all-too-tempting-to-ignore topic. (On that note, special thanks go to the new breed of climate activists who appear to have finally cracked the code on communicating climate as a systemic challenge, and as a social justice and equity issue.)

Finally, a note of remembrance and gratitude for my Dad, Mike, a publisher who instilled in me a love of books, a love of life, a love of people, and a sense that we should do what is right, preferably while not taking ourselves too seriously.

I would so dearly have loved for you to hold this book in your hands.

Preface:
The Night I Went Drinking
and the World Fell Apart

"Why are there ads for professional-grade respirators and hand sanitizer all over our Amazon account?"

I was barely awake when my wife walked into our bedroom. She was puzzled at what seemed to her to be a change in Amazon's algorithms.

This was February 25, 2020. I had been out to Dain's Place (my local dive bar) the night before with two friends of mine. In passing, I had mentioned the unprecedented lockdown that was currently underway in Wuhan, China. A novel coronavirus had recently been identified, and hospitals were swamped with patients reporting respiratory issues. I pondered aloud whether the world might be overreacting to what, some were saying, could just be a bad case of the flu.

My friends—both more scientifically literate than myself—raised their eyebrows. This, they told me, was an extremely dangerous situation. One of my drinking companions was building a tech startup focused on scientific research data. He explained to me that we should fully expect to see the infection rate skyrocket as the virus began to spread. He pointed me to credible sources suggesting that without effective mitigation efforts, as much as 40–70 percent of the world's population could get infected. In a scenario like this, he said, the deaths would likely be in the millions.

Then the conversation moved on.

We drank some more beer. We talked about politics and bad movies. Then we drank even more beer. Eventually, we said our good-byes. (My friend looked at me like I was insane when I went to shake his hand.) Meandering home from the bar, pleasantly drunk, I began to reflect on their warnings. Then, slowly, I started to freak out.

And that's the story of how I came to be sitting up at midnight, searching for face masks and respirators on my wife's Amazon account.

A Gradual Social Reckoning

Of course, we now know that this was the early days of the COVID-19 pandemic. This was a catastrophic global event which, at the time of writing, is still ongoing. And I believe it offers some lessons for us about how we—as individual citizens—can engage with a dangerous crisis that is more complex than any single one of us can tackle alone.

Much like with the topic of climate change, in the US at least, not everyone was listening to the science during the early part of the crisis. It was almost as if there was an invisible split between those of us who had been clued in to the impending disaster, and everyone else. Thanks to nothing more than a chance conversation with a couple of friends, I now found myself on the side of the forewarned. Writing on Twitter, *Atlantic* staff writer Amanda Mull opined on the strange social dynamic that many people found themselves in:

> [I] feel like every social circle right now has someone who has read all the coronavirus news out of China and Korea and Italy and knows what's coming and they're in charge of gently explaining things to all their friends who heard "a bad flu" three weeks ago and checked out.

Gradually, however, reality caught up.

News from Italy was particularly grim. Hospitals were overwhelmed, the death rate was spiking, and eventually the whole of the country, and many neighboring countries too, had to go into complete lockdown. Serious epidemiologists began appearing in the media, suggesting that the US was only a few weeks behind.

Action Is Contagious Too

In the absence of federal leadership, local, regional, and civic entities stepped up. The NBA announced that it was canceling all games on March 11. In the days that followed, cruise lines and sports leagues, museums and concert halls, school systems and conference centers made similar announcements. And the number of companies sending employees to work from home snowballed into the thousands.

Individuals began canceling events too. One friend of mine went into self-quarantine with her family due to pre-existing health issues that made them high risk. Within a matter of days, states like California and New York began issuing orders for partial or complete lockdowns and shelter-in-place advisories. Before long, the vast majority of the country was in some form of state-mandated social distancing, including my adopted home state of North Carolina.

In other words, steps that had seemed extreme began to feel normal. What had felt like fear mongering began to feel like an expression of our civic duty. And each action by one entity made the next, more ambitious action by another significantly more socially and politically acceptable.

Individuals and institutions started by doing what they could—often imperfectly, and often alone. But it took hold because through their actions, however incremental, those individuals were able to bring others along with them. Once momentum reached a critical mass, it all happened incredibly quickly. Urgent, sustained, and diverse pressure from within the system was able to change conversations and build something that—temporarily at least—approximated a consensus. (Sadly, that consensus didn't last. But that's another story…)

The parallels with the climate movement are painfully obvious. We too have found ourselves warning of a dire threat that the system is not taking seriously enough. We too have been asking ourselves where and how we can take action on a crisis that is bigger than any one of us. And we too must accept that whatever we do, however far we go in our own lives, it will mean next to nothing unless we can bring others—meaning entire societies—along for the ride.

And that means keeping our eye on the big picture.

Getting to the Point

It's customary for any book on the climate crisis to spend at least one chapter laying out the dire nature of the science. I'm going to skip that formality.

It's not that the science doesn't matter. (It does.) And it's not that the situation isn't serious. (It absolutely bloody is.) It's just that I assume that anyone reading this book is—to borrow a phrase from my earlier coronavirus analogy—already on the side of the forewarned. If you are, for whatever reason, new to the topic and would like to learn more, or if you simply enjoy reading books about terrifying planetary emergencies, then please take a look at the resource list at the end of this book. It includes both overviews of the crisis itself, as well as more detailed resources on what we, as a society and as individuals, can and should be doing about it.

But for now, for this book, the topic is not the crisis itself—that's the water we are swimming in. Instead, it's an attempt at asking ourselves what it looks like to live within that crisis, and to play a productive part in moving us toward solutions, even as we accept that we are also part of the problem.

It's partly a recounting of my own experiences. I've been aware of the climate crisis, and active in trying to help solve it, since my teens. Yet I now, miraculously, find myself in my forties with very little progress to show for my efforts, and with many of the trappings of a comfortable, middle-class, Western existence, not to mention the carbon footprint to go with it.

The rest of this book is the result of countless conversations with activists, academics, writers, and engaged citizens from many walks of life, who are each trying to find their own place within this complex, confounding fight for survival. To all of them, I owe an immense debt of gratitude. Some of the voices you'll hear are avid champions of transforming our own lives and leading by example. Others are dismissive of individual carbon footprints, and more focused on holding the powerful to account. But all of them share a similar end goal—the complete transformation of our society.

Ultimately, they have more in common than they have that divides them. My hope is that we can all move forward. Together.

1

We're All Climate Hypocrites Now

"You can't be an environmentalist and eat meat,"
says the vegan as he steps onto the plane.

"You can't possibly live sustainably and drive a car,"
says the cyclist as she tucks into a burger.

"You can't be green and not compost,"
says the gardener as they plan an extension to their house.

Anyone who has been involved with modern, mainstream environmentalism will be familiar with the selectively applied purity test. Sometimes the gatekeeping is explicit, and sometimes it's implied. Sometimes we even imagine it as coming from people who have no intention of judging us at all, but who are simply doing a better job than we are at reducing their own environmental impact. Whatever the delivery mechanism, it's become so pervasive that it has shaped the conscience of those who don't really consider themselves environmentalists at all. Given the market demographics of folks who read books like this, you are most likely familiar with the predicament that such framing can cause.

On the one hand, you know that we are in the midst of perhaps the worst crisis humanity has ever faced, and you are rightly concerned. On the other hand, you are likely spewing significant amounts of carbon dioxide into the atmosphere every single day.

Indeed, while my fantastic publisher does what they can, production and distribution of this book itself—whether you are reading it in electronic or print form—is only made possible by the consumption of fossil fuels and other natural resources.

So what's a concerned global citizen supposed to do?

For all the great work being done around the world, there is a basic assumption pushed by the dominant culture that a person's contribution to the climate fight is largely, if not exclusively, measured by their personal carbon footprint. Say one word about the climate crisis, or the need to divest from fossil fuels, and you'll soon be met with a question about how you traveled to work today, or where the electricity powering your computer comes from. Even if you are just beginning to learn about the issue, there's a good likelihood that you've received more advice on changing your diet, refusing straws, or abandoning consumerism than you have on activism, advocacy, or organizing. In other words, you've been told how *not to contribute* so much to the problem, but not necessarily how you can be *most effective in actually fixing it.*

In many ways, the problem stems from a logical reading of the crisis we are in. Whether through driving or flying, eating a burger or streaming movies, we—meaning those of us enjoying at least a moderate level of material comfort—*are* all contributing to the climate crisis. And if it's our daily lifestyle choices that make us a part of the problem, then maybe we just need to make better choices. After all, if we don't put our own house in order first, aren't we basically just climate hypocrites if we start pointing the finger at the Koch Brothers or Exxon Mobil?

Well, it all depends on what you mean by "hypocrite."

What Does 'Hypocrite' Even Mean?

In a fascinating paper published in the journal *Frontiers in Communication*, a team led by Shane Gunster of Simon Fraser University in Canada looked at how terms such as "hypocrite" and "hypocrisy" show up in coverage of the climate crisis. Analyzing op-eds from both conservative and liberal newspapers around the world, what they found was a remarkable diversity of uses. On the one hand, opponents of climate action would often use allegations of hypocrisy as a cudgel to undermine celebrity activists and "elite" environmentalists whose ideas they opposed. Over time, such arguments have created a tricky dilemma for the climate movement:

> This one-dimensional but compelling equation of environmentalism with sacrifice leaves climate advocates in a proverbial no-win situation when it comes to reconciling behavior with beliefs. As author Lynas (2007) wryly observed in a Guardian op-ed, "climate activists I know who do walk the walk (eschewing all flights, for example) look prim and obsessive, as if they are out of touch with the concerns and pressures faced by ordinary people." Yet the views of those who do resemble "ordinary people," and therefore fail to pay adequate behavioral homage to the gravity of the crisis, are likewise subject to ridicule and dismissal.

However, opponents of climate action aren't the only ones engaging with the topic of hypocrisy. Gunster and his team also found plenty of articles from pro-environment voices too, many of them exploring the all-too-familiar gap between activists' professed values and their everyday behaviors:

> The most interesting and provocative explorations of climate hypocrisy were those which simultaneously accepted the claim that individuals do bear (some) responsibility for their carbon-intensive behavior (rather than simply deflect such claims to structures and institutions) but then challenged the assumption that such responsibility is best (and solely) discharged through consumer action.[1]

It is these, more nuanced discussions of hypocrisy that I believe offer us a path forward. And they do so by pointing to one of the biggest fallacies of our culture.

Rational Choice Is No Choice At All

Economists and politicians have been mythologizing the "will of the market" as a mysterious, all-knowing force for years. Yet the supposedly rational choices we make are heavily influenced—if not quite predetermined—by factors that are way outside of our own individual control. As my friend David Monje, a teaching associate professor of Media and Technology at the University of North Carolina, put it to me when we were discussing an early draft of this book, "Rational choice theory is really, really stupid."

From taxes to planning laws, and from government subsidies to cultural norms, our society makes certain behaviors easy, cheap, and socially acceptable. Meanwhile, it makes other behaviors so expensive and onerously difficult that only the hardest of the hardcore among us can even hope to stay on the straight and narrow. Sure, each of us plays a role in setting these systems and norms. It's undeniable, however, that some forces—and some entities—play a larger role than others.

Unless we acknowledge and seek to change the hidden ways that society shapes our decisions, then focusing the discussion primarily on the choices that each of us make in our daily lives is not just ineffective, it's potentially downright counterproductive.

Undermining the Messenger

Perhaps nobody has had their work more fundamentally undermined by our culture's limited, individualistic framing of environmentalism than former Vice President Al Gore. When his documentary, *An Inconvenient Truth*, premiered in 2006, it dramatically raised awareness of the climate crisis and brought in $49.8 million at the box office. Using little more than a PowerPoint presentation, the film introduced climate science to a mainstream audience. Yet rather than grapple with the complex, terrifying facts presented in the film, critics were quick to change the subject.

One free-market think tank, for example, released a report claiming that Mr. Gore's house used 20 times more energy than the average American family home. And while Al Gore's spokespeople responded with statistics about his carbon offsets and energy-efficient renovations, the distraction campaign had already worked. Discussion had shifted from the systemic underpinnings of our reliance on fossil fuels and was instead now focused on the personal choices of one specific individual.

"Al Gore's Inconvenient Truth: a $30,000 Energy Bill," cried one particularly snarky headline from Jake Tapper over at ABC News.[2]

Shaming activists for what they are not doing has proven to be devastatingly effective. Not only does it undermine the credibility of the immediate target, but it simultaneously redirects the focus away from the societal-level solutions that could bring about change at the scale and pace that's necessary.

It also sets the bar almost impossibly high for others who would like to join the movement. How can I, as an individual, demand an end to fossil fuels if I still rely on them to get me to and from work? Who am I to question subsidies for airlines, if I'm still flying to see the family at Christmas? Yet if we pause for a moment to consider how we talk about other societal problems, it becomes easier to see that the basic premise of such supposed hypocrisy is self-defeating bullshit.

If a citizen were to advocate for higher taxes on cigarettes, for example, it would hardly undermine their argument to reveal that they themselves were addicted to nicotine. In fact, it would be one more proof point among many that we can't rely on voluntary abstinence in the face of a socially harmful and manipulative industry.

To cite another example, if a billionaire were to campaign for higher taxes on the rich—as some enlightened "One Percenters" have actually begun to do—then their case is strengthened, not weakened, by the fact that they can't fix poverty through their own individual acts of philanthropy. The fact that they find it necessary to advocate for structural changes—changes that would directly harm their own narrow financial interests—demonstrates the systemic nature of the problem they are seeking to fix.

I'm not going so far as to suggest that it makes you a *better* advocate for climate justice to fly or own a big house. Neither am I arguing *against* lower carbon lifestyle choices as one strategy among many we can deploy. I am, however, saying that a high personal carbon footprint shouldn't preclude you from doing your part. And I'm urging you to focus your energies on where you—personally—can have the biggest impact on the structures and society around you. Not only will adopting a more systemic perspective help you to prioritize your efforts. It will also, I believe, undermine a key strategy of those who would hold back progress.

A Convenient Mistruth

In 2007, the publishers of the Oxford English Dictionary chose "carbon footprint" as their UK Word of the Year. (In the US, the honor went to the not entirely unrelated term "locavore.") Since then, the concept has embedded itself so integrally in the climate debate that it can actually be hard to remember just how new it is, or who helped to elevate it in our popular consciousness in the first place: namely none other than your friendly, planet-warming oil conglomerate BP.

While the term itself had been floating around in academia for some time,[3] one of the first publicly available online carbon footprint calculators was promoted and popularized as part of BP's highly problematic "Beyond Petroleum" rebranding campaign that it attempted in the mid 2000s. Not long after, the world watched in horror as the Deepwater Horizon drilling rig went up in flames, pouring 4.9 million gallons of oil into the Gulf of Mexico in the process.

As Mark Kaufman of Mashable has documented, BP's championing of carbon footprints should be viewed not simply as a naïve or imperfect effort at corporate responsibility, but rather as a direct and calculated attempt to shape discussion of the problem in BP's favor:

> One of the creators of BP's ad campaign who approached Londoners on the street, the PR professional John Kenney, later acknowledged it was all a marketing scheme, not a sincere effort to promote BP's low-carbon or renewable energy transformation.

"I guess, looking at it now, 'beyond petroleum' is just advertising," Kenney wrote in a New York Times Op-Ed in 2006.

"It's become mere marketing—perhaps it always was—instead of a genuine attempt to engage the public in the debate or a corporate rallying cry to change the paradigm." BP, powerful and wealthy, signaled it would wean itself from oil. "Only they didn't go beyond petroleum," wrote Kenney. "They are petroleum."[4]

Contrary to popular opinion, oil companies have demonstrated time and again that they are actually all too happy to talk about the climate crisis. They just want you to know that it's mostly your fault.

Yet fossil fuel interests aren't the only people playing into this narrative.

Eco-Moralism Runs Deep

George Monbiot, a British environmentalist and writer, has become a household name for his unflinching writing about the climate emergency. While much of his focus has been on the structural underpinnings of the problem, Monbiot has also directed his rhetorical fire at his eco-minded peers.

From meat eating to outdoor patio heaters, many trappings of the modern, middle-class existence have fallen afoul of Monbiot's criticism. But his willingness to call out double standards is perhaps best demonstrated by a 2006 article for *Alternet*, in which he made the moral case for curbing aviation:

If we want to stop the planet from cooking, we will simply have to stop traveling at the kind of speeds that planes permit. This is now broadly understood by almost everyone I meet. But it has had no impact whatever on their behavior. When I challenge my friends about their planned weekend in Rome or their holiday in Florida, they respond with a strange, distant smile and avert their eyes. They just want to enjoy themselves. Who am I to spoil their fun? The moral dissonance is deafening.[5]

To be fair, the core point that Monbiot is making is hard to refute. If we really are in an existential climate crisis (we are), if we really need to cut carbon emissions as quickly as possible (we do), and if millions of people will die if we fail (they will), then taking a vacation closer to home would seem like a small price to pay for safeguarding civilization from catastrophe.

Yet it never seems that simple when it's us making that choice. I know this from personal experience.

Nothing's Ever Easy

In the Spring of 2005, I made the decision to give up flying.

I was 27 years old, living in my native South West England, and working as director of sustainability for an independent academic publishing house that my parents had founded several decades before. I had used that position to advance my green agenda, including enacting company policies that restricted corporate flying and encouraged train travel instead. However, I had recently taken part in an experiment to calculate my own carbon footprint, which back then was a relatively novel idea. In doing so, I had been profoundly shocked by the impact of my conference travel. Despite all my best efforts to car pool to work, eat a mostly vegetarian diet, compost, buy second hand, and generally behave like a good green activist, the job that I had chosen to do (and/or the family I had been born into) meant that I was causing nearly two times as much damage to our atmosphere each year than the average British citizen.

After much soul searching, I decided that I could no longer justify flying so much. I tendered my resignation from the family business, and started making plans for a career move into environmental education. First, however, I had to take one last flight across the Atlantic. (Ironically enough, this was for a conference on "sustainable" tourism.) While there, I visited Edward, my old college roommate who was now living in Carrboro, North Carolina. And because he was busy, I ended up going out for drinks with Jenni—a friend of his whom I had met a few times before.

A few hours and several beers later, we—or at least I—promptly fell in love. And thus began a long-distance relationship that changed my plans forever. Within the space of a year we were engaged, I was applying for US residency, and resigning myself to a lifetime of transatlantic flights to see my family, and to recharge on the real cask ales of my homeland.

I don't share this story to make excuses, nor to seek pity. (There is an almost laughable amount of privilege involved in claiming this as a personal problem.) Nor am I arguing that flying less is pointless. In fact, I greatly admire those who have given up flying. But I do share this story to simply make the case that any attempt to promote greener lifestyle choices can and must accept that we are all starting from different places. What's easy or rewarding for one person may be difficult or repulsive for another. What's exciting and aspirational for one demographic might be too expensive or elitist for another. Choosing not to fly may actually mean fantastic rail travel adventures, or more time at home, for some. For others, however, it may mean compromising your career, disappointing family and loved ones or, as in my case, never visiting your parents or drinking proper beer again.

To be fair, even the most adamant "No Fly" advocates are aware of this challenge. In another particularly moving article for *The Guardian*, George Monbiot—whom I quoted earlier pointing fingers—acknowledged the challenge for international families, referring to a concept that he describes as "love miles":

> If your sister-in-law is getting married in Buenos Aires, it is both immoral to travel there, because of climate change, and immoral not to, because of the offence it causes. In that decision we find two valid moral codes in irreconcilable antagonism.[6]

According to Monbiot, the logical conclusion of this "irreconcilable antagonism" would be to curtail all non-essential (i.e., non-love related) air travel. That would mean an end to weekends in Ibiza, or Brits shopping in New York. It would mean conducting business meetings via video conference, and making transcontinental journeys by train. And it would mean that journeys around the world would be reserved for visiting people we hold dear. Even then, he predicted, it would involve "both slow travel and the saving up of carbon rations."

Whether or not aviation will need to be curtailed to exactly that degree, or through those specific methods, is not for me to say. Monbiot's proposed solution does, however, point to how the issue will be solved. And that's systemically.

The Limits of Personal Responsibility

Here's the uncomfortable and inconvenient truth: the vast majority of environmentalists, myself included, are doing more than most of our peers. I've gone to great effort to install insulation in our 1936 home, for example. I drive an old, used electric car. I often ride an e-bike around town. And I've cut back on my meat eating considerably. Yet an unfortunate mix of societal influences, the car-centric sprawl of the region I now call home, my socioeconomic status, and my own all-too-consumerist failings mean that I am doing only a fair-to-middling job at actually cutting my emissions. The last time I calculated, my footprint was only some 25–30 percent lower than the US average. (Meaning it's also several orders of magnitude higher than your average global citizen.) I would hazard a guess that many of my eco-minded peers are experiencing similarly mediocre results.

On a related note, environmental justice advocates will rightly point out that there are huge inequities in carbon footprints depending on wealth and income, and that curtailing the impact of high emissions lifestyles is clearly a moral imperative. Yet in high wealth countries, even folks who are living in poverty or on low incomes will still have outsized carbon footprints when compared on a global scale.

And these carbon footprints exist because of factors well outside of any individual citizen's control.

When this book asserts that 'we' are all climate hypocrites, I do not intend it as an attack on any one of us. Nor do I mean that either 'blame' or 'responsibility' is equally shared. I simply make the case that living sustainably in our current system is nearly impossible. We may therefore want to be careful about pushing for purity, or feeding an individualistic narrative that ultimately perpetuates the status quo. Instead, I suggest, we should identify paths for effective, mass mobilization that makes better, more equitable, and more rewarding low carbon living an attainable reality for all of us.

Why Individual Action Still Matters

When I first started working on this book, I intended to debunk the idea that individual action was central to creating change at all. I made a note to myself about what I thought was a useful analogy: the transatlantic slave trade didn't end because people stopped eating sugar. Yet it turns out that this is only half true.

In fact, sugar and rum boycotts were a key strategy of abolitionist groups. At one point, some 400,000 people in Britain alone were said to be boycotting slave-grown sugar. As part of that effort, James Wright, a merchant from Suffolk, took out an advertisement in the newspaper:

> ...I take this method of informing my customer that I mean to discontinue selling the article of sugar when I have disposed of the stock I have on hand, till I can procure it through channels less contaminated, more unconnected with slavery, less polluted with human blood....[7]

Contrary to my ill-informed assumptions, boycotts were actually pivotal in shifting the political dynamics of slavery. They helped make the moral case for abolition, they gave individual citizens a tangible way to live their values in their daily lives, and they exerted a direct economic pressure on the powerful forces that were profiting from business-as-usual.

Yet the abolitionists promoting boycotts weren't suggesting that nationwide abstinence from sugar was the ultimate solution to ending slavery. Instead, they were *tactically* pulling the lever of abstinence with a *specific* end goal in mind, and they were doing so as part of a *broader set of strategies*. Boycotts, by themselves, were never going to be enough to bring this murderous industry to its knees. They were, however, an accessible entry point for would-be abolitionists to flex their muscles. (This was especially true for women, who didn't yet have the vote, but who did have a say in their personal household purchases.)

The lesson for those of us trying to mobilize on climate is not to ignore questions about what we should or shouldn't be doing in our personal lives. Rather, it's to rethink *why* those actions matter. In his book *There Is No Planet B*, author and environmental activist Mike Berners-Lee defines the challenge like this:

> We need to think beyond the immediate and direct effect of our actions and ask more about the ripples they send out...[8]

I couldn't agree more. First, however, we need to talk about breakfast.

2

Wants and Needs

I really want a crepe pan. But not just any old crepe pan.

I want one of those fancy French ones that are almost impossibly large, entirely flat, with just a teeny-tiny rim around the edge. Ideally it would come with a wooden crepe rake—a tool that helps you spread the batter into a consistent, paper-thin circle.

With it, I am 100 percent sure I could create something akin to those beautiful, smooth yet lacy crepes the French make. And I am also convinced that my daughters would love me just a little bit more than they already do.

The one I want is regularly advertised in an e-newsletter I get from a fancy kitchenware site called *Food 52*. This company is the kind of trendy, online retailer that also pushes single-origin olive oils and hand-carved chopping boards. Everything is more expensive than it probably needs to be, and is photographed in beautiful, Instagram-ready designer kitchens.

Every time I receive one of those newsletters, I fantasize about how different my morning routine would be if I only had this pan. At least, that's what I tell myself while scrolling through the product offerings for the umpteenth time. Here's the thing though: I really don't need one.

We make crepes about five times a year. I have a perfectly good non-stick omelet pan. With that pan I turn out delicious (if I say so myself) crepes that my children will inhale in an instant while my wife—a registered dietitian—looks on disapprovingly. In fact, there is a solid argument to be made that a crepe pan would actually be a detriment to our lives. Not only would it cost money and sit mostly idle, but it would also take up storage room that we could put to much better use.

So far, at least, I have resisted the urge to purchase one. But there's no guarantee that my resistance will last forever. After all, I really want one.

Voting and Shopping Are Not the Same Thing

There's a popular axiom among a certain breed of environmentalist, used often by the movement of Certified B Corporations of which I am an enthusiastic participant (see Chapter 7). It urges us to "vote with our dollars." In other words, by choosing products and services that align with our values, the argument goes, we can create the world we want to see, just as we might vote for the political vision of a candidate whom we admire.

There is some power in this idea. After all, we get to vote at the ballot box only once every few years. Yet most of us shop for something at least several times a week. If our consumer purchases have both positive and negative impacts (and they do), then directing those choices toward greener, lower-carbon products and services has the potential to have a positive influence on society.

Yet equating voting to shopping has always gotten on my nerves.

Comparing my choice of which product, retailer, or producer I choose to the act of casting my vote—a right which people have fought and died for—is misleading at best, if not borderline offensive. Not only does it cheapen the act of voting, but by placing the primary burden of social change on the "free" market, we are implicitly surrendering the ideals of equity.

If we center our strategy for change on shopping behaviors, then those who have more money will have more influence too. "One person, one vote" all too easily becomes "one dollar, one vote." This is hardly a recipe for empowering those whom our system has marginalized. Leaving these moral questions aside for a second, there is also a more practical concern. That's the fact that when we are acting and thinking as consumers we are not exactly in our wisest or most altruistic state of mind.

The Irrational Consumer

Dan Ariely, a professor of behavioral economics at Duke University, has spent years researching our consumer impulses. And if there is one thread that runs throughout his research, it's the fact that our behavior is often governed by influences that are not entirely rational. Among his many findings:

- If you ask a group of people to tell you the last two digits of their social security number (SSN), and then convert it into a dollar amount (e.g. XXX-XX-XX78 = $78), and if you then ask them to participate in an auction, the participants with the numerically higher last digits will be significantly more likely to spend larger amounts of money.
- A consumer's perception of how valuable an item is can be greatly shaped not just by the item itself, but by the value of the items around it. This means a retailer can skew how much you are willing to spend on a product simply by placing a highly priced product right next to it.
- We are more likely to value a three-dollar latte than we are three actual dollars in cash, even though the dollars themselves are obviously more useful to us in terms of flexibility.[1]

None of these insights are directly related to carbon emissions. Yet they are important because they teach us one very simple thing: Our consumer brain is not very smart. More precisely, it is not built for making complex, objective, or strategic decisions. Whether it's me desiring that crepe pan I know I don't need, or a participant in Ariely's experiments who is willing to spend more because they just talked about their social security number, we are influenced by difficult-to-control impulses that are at their strongest when we are reaching for our debit cards.

And yet—as we will see throughout this book—the discourse around the climate crisis has too often identified both the problems and the solutions as little more than the sum of our collective consumer choices. (This holds true, even when the debate is about what *not* to buy.)

Behavior Is About Design

In the book *Hacking Human Nature for Good*, Ariely, together with his co-authors Jason Hreha and Kristen Berman, actually tackled some of the reasons why encouraging climate-friendly behaviors is such a tough challenge. These reasons include:

1. **There's friction:** Meaning it's often easier to do nothing, or do the wrong thing, (e.g., leaving the lights on or not adjusting our thermostats) than it is to do what is right (e.g., turning off the lights or turning down the AC).

2. **The pain of acting now overshadows delayed benefits:** Meaning we want what we want when we want it (remember the crepe pan?). Meanwhile the benefits of environmentally preferable choices might not be felt for many years, perhaps even generations, to come.

3. **People don't think about the benefits at the right time:** We might worry about climate change when we are walking in the forest, reading the Sunday papers, listening to a sermon or watching a documentary. Yet as we stroll the aisles of the supermarket, we're focused on the rumbling in our belly and the pressure to feed the kids—and we're in an environment that has been designed to actively accentuate our consumer urges.

4. **People don't agree it's a good idea:** While growing numbers of people are now waking up to the threat of climate change, some are still in denial.

In short, we humans are wired in such a way that it puts any form of long-term decision making at a distinct disadvantage when we are going about our daily lives. This holds especially true when we are exercising our consumer brain. And while it might be tempting to assume that education will help bring about change, it's not quite as simple as just letting people know that our planet is in trouble. Again, Ariely and his co-authors explain why:

> Knowledge is about tomorrow. In the now, we're driven by the environment we currently live in. The major theme, and arguably the biggest principle within behavioral economics, is that environment determines our behavior to a large degree, and to a larger degree than we intuitively predict.[2]

While it might be depressing to think that knowledge alone is not enough to inspire us into action, Ariely and his team argue that there are actually design and engineering workarounds to almost every barrier they identify.

Smart thermostats, for example, can help eliminate the friction involved in remembering to adjust your heating or cooling. Meanwhile Dartmouth College has experimented with linking energy-saving behaviors in student dorms—for example, turning out lights—to a real-time display that depicts the health and well-being of an animated polar bear. The results of that experiment showed that you can significantly incentivize conservation. (This was especially true when positive results were correlated with rewards.)

It's certainly encouraging to know that product design, marketing or environmental engineering can be used to "hack" human behaviors. The more we create environments and social structures that encourage lower-carbon living, the less we have to worry about winning over individual hearts and minds. If you're an architect, a product designer, a city planner, or a college administrator, there is every reason to learn from these strategies and apply them to your sphere of influence.

Yet however important these insights are, it's also possible to reach another, perhaps even more important, conclusion from Ariely and his team's work. Not only do we need to rethink *how* we approach behavior change on the individual level but, if changing that behavior requires such detailed engineering and design, then we might want to rethink *how much* we focus our attention on the individual scale at all.

Maybe we need to think bigger.

The Roles We Play

In November of 2013, writer, teacher, and activist Micah Bales was reading an article about the then relatively new Affordable Care Act (aka Obamacare). As he later recounted in a blog post for *Sojourners*,[3] he was troubled by the very premise of how the topic was being framed:

> The dilemma was presented as a story of tension between healthier consumers and less healthy consumers fighting to get the best deal for their health-care dollars. But could there be another way of thinking about health care, and about our society as a whole? Is there a framework that would allow us to consider these questions in a way that assumed connection, caring, and community between individuals, rather than the zero-sum competition of the market?

The problem, he argued, was that we have come to rely far too strongly on the idea of individuals first and foremost as consumers. Instead, we might want to think of a more expansive and connected idea of where our true purpose lies:

> The idea of citizenship could offer a positive antidote to the consumeristic worldview. While consumers have only unmet desires and (hopefully) means to pay for it, citizens have rights, responsibilities, and a role within a larger community. What might change if we thought in terms of rights and responsibilities, rather than in terms of consumer desire and spending? In short, what would be the effect of a worldview that is primarily civic rather than hedonistic?

Bales didn't stop with the idea of citizenship, however. He also argued that we need to reexamine another problematic aspect of Western culture(s):

> Yet, there are definitely problems that this worldview based in citizenship would fail to address—in particular, our culture's unbalanced focus on the individual. Even as a nation of citizens, it would still be easy for us to think in terms of my personal rights and my personal responsibilities. We would no longer be hedonists, perhaps, but we would still be individualists.

This idea that even a citizen mindset will fall short without also returning to community, has important implications for the climate movement in particular. While there are many of us climate activists who reject the label of consumer, and would much rather think of ourselves as citizens, the magnetic pull of individualism still plays a strong role in how we envision our roles within the movement. It also limits our vision of what we could achieve together.

Abstinence Is Still Individualism

So far, my complaints about "lifestyle activism" have focused primarily on the idea of ethical consumerism. To be fair, however, many of the most impressive proponents of behavior change aren't really arguing for shopping as a path to solving this crisis at all. In fact, more often than not, they are arguing for quite the opposite: a radical reduction in how much we shop in the first place.

This approach makes a lot more sense. If we want to diminish the nefarious power of the oil producers and car companies, airlines and corporate retailers that are endangering our future, then refusing to hand over our dollars to them is one sure-fire way to minimize their influence.

Indeed, I too take steps to limit my consumption. Yet I don't believe it should be the central strategy we rest our hopes on. Not only, as we have seen, are we asking people to overcome some pretty powerful and irrational urges, but we are also still centering our conversation on tactics that are primarily individualistic. As Micah Bales argued in relation to the Affordable Care Act, we can't simply move from being individual consumers to being individual citizens. We have to start seeing ourselves as one part of a much larger and more complex whole.

Finding a Bigger Political Canvas

In June 2015, author and social activist Naomi Klein gave a commencement address at the College of the Atlantic. Given the audience were all attendees at a liberal arts college that focuses on human ecology, Klein didn't waste too much time trying to motivate them to pursue social change. Instead, she assumed they were already on board, and delved straight into *how* they could be as impactful as possible in driving that change.

As part of that address, she recalled some of the experiences she had when she was researching her first book, *No Logo*. That book tackled the impact of global supply chains on workers in lower-income countries. Referencing discussions with labor organizers in the Philippines and Indonesia, she recalled that many of the activists she spoke with were utterly baffled when she brought up the topic of ethical consumerism.

Having stood together in the face of corporate intimidation, and sometimes actually won, they simply couldn't fathom what a person's choice in sneakers had to do with holding the powerful to account. Yet back home, Klein said, even those who understood the devastating impacts of sweatshops were all too quick to shift the conversation to which brand of shoe they should buy instead:

> The irony is that people with relatively little power tend to understand this far better than those with a great deal more power. The workers I met in Indonesia and the Philippines knew all too well that governments and corporations did not value their voice or even their lives as individuals. And because of this, they were driven to act not only together, but on a rather large political canvas.[4]

Interestingly, the modern environmental movement hasn't always been so focused on the individual. In fact, it achieved considerable success pursuing exactly the kinds of movement building and public pressure campaigns that these organizers in Indonesia and the Philippines would recognize. Somewhere along the way, however, it lost its way. And it's important to trace that journey if we want to figure out where we go next.

3

How "Green" Lost Its Groove

In an article that spread like wildfire around the enviro-leaning corners of the Twittersphere, climate essayist and podcaster Mary Annaïse Heglar lamented how a focus on moral purity has held back the green movement for years. Provocatively titled "I work in the environmental movement. I don't care if you recycle." the essay described an awkward encounter at a friend's birthday party.

Having just introduced herself to another partygoer, Heglar recounted how the man learned that she worked for an environmental organization, and then proceeded to confess a litany of supposed eco-sins. From taking a cab to eating meat, the unnamed guest seemed to be unburdening himself of his guilt. It was as if, writes Heglar, he had decided she were "some sort of eco-nun."

When she tried to reassure her new acquaintance that she didn't blame him for his actions, but rather blamed the system for how it shaped his decisions, he pivoted to an entirely different and equally unhelpful perspective: there's no point in even trying to create change anymore because we've already doomed ourselves to environmental disaster.

> Sadly, I get this reaction a lot. One word about my five years at the Natural Resources Defense Council, or my work in the climate justice movement broadly, and I'm bombarded with pious admissions of environmental transgressions or nihilistic throwing up of hands. One extreme or the other.[1]

As Heglar argues, neither of these extremes is likely to build the kind of movement that's capable of taking on Exxon Mobil. In a separate interview with Yessenia Funes for *Atmos*, Mary points to the overwhelming whiteness of mainstream environmentalism as one source of this individualistic worldview:

> When I first got into climate work, it was like, OK, well, these narratives feel very shallow. They feel like they're coming from people who have never had to fight for their lives before, which I just can't relate to. I felt like I was with a bunch of people who didn't know what they were doing.

But how, given the obvious threat that climate change and mass extinction pose, did we find ourselves at this impasse?

Well, it's a long and somewhat depressing story.

Dilution of a Movement

Ask experts when the modern environmental movement began, and most will point to the first Earth Day as a pivotal moment. Inspired by the public outcry around pollution in the '60s, and conceived of as a teach-in by Senator Gaylord Nelson, the first nationwide event attracted around 20 million participants. Specific actions included marches, rallies, tree plantings, and more. Inherently political, the original Earth Day included events focused on such forward-thinking themes as environmental racism and was, and still is, considered to be one of the largest demonstrations in US history.

It worked, too. In the years that followed, politicians from both sides of the aisle responded to the pressure they were feeling from the public. They passed significant legislation that included the Clean Water Act and the Endangered Species Act. Yet the momentum eventually fizzled out. As Adam Rome, author of *The Genius of Earth Day*, told Kate Yoder of *Grist*,[2] the Reagan years eroded the public's faith in the power of government. They also blunted people's belief that citizens' movements could really bring about any form of meaningful change. This shift in how we thought about our roles and responsibilities in a democracy contributed directly to the hollowing out of Earth Day. In fact, Rome argued that we can pinpoint 1990 as the specific year that Earth Day became something entirely different than it was originally intended to be:

> "It's weird to be able to be that precise about it," Rome said. In the 1980s, the Reagan administration, in its zeal to unleash the free market and roll back regulations, managed to erode public trust in the government and faith in collective action, Rome said. "That pushed a lot of people to say, 'If we're not going to get new legislation, how do we get change?'"

The resulting rethink spawned a shift away from political and legislative strategies, says Yoder. Instead, activists gravitated towards personal responsibility and lifestyle change as a place where they still had agency. A popular pamphlet, which sold more than five million copies, promoted 50 *Simple Things You Can Do to Save the Earth*. From there, it was but a short, socially responsible bike ride to a world in which environmentalism largely became synonymous with ethical consumerism, personal responsibility, and being a "good steward" of our planetary resources.

It was a framework that would shape the discourse around climate for many decades to come.

A Missed Opportunity

It's a black and white scene. There's a couch on the pavement in front of the US Senate. On that couch sit Nancy Pelosi, Democratic speaker of the House, and Newt Gingrich, Republican former speaker of the House. The pair look straight into the camera and declare that, while they don't always agree on everything, they do both believe that America can and must take action on climate change.

"Together, we can do this." declares Pelosi.

Released in 2008, this televised ad was part of the "We Can Solve It" campaign sponsored by Al Gore's Alliance for Climate

Protection. While it seems like a quaint, even surreal throwback now—especially given Gingrich's decidedly mixed record on science and the environment— it does represent a brief moment in time when America might have started taking climate change seriously. (Gingrich did later call the ad "the dumbest single thing I've done."[3])

This was not too long after the release of An Inconvenient Truth, and for those of us who had been campaigning for climate action for years, there was a palpable sense that the world was finally beginning to take notice. Yet for all the talk of bipartisan consensus that followed the release, it's hard not to look back now and suspect that the opportunity was always going to be squandered.

For while An Inconvenient Truth clearly and urgently laid out the dangers and causes of the climate crisis, and campaigns like "We Can Solve It" appeared to bring people together to tackle it, much of this apparent awakening was still looking at the crisis through a lens of equal culpability and incessant individualism. In the book that accompanied An Inconvenient Truth, for example, the section on solutions was buried at the back like an afterthought. And the message of that section was pretty unambiguous when it came to assigning responsibility:

> When considering a problem as vast as global warming, it's easy to feel overwhelmed and powerless—skeptical that individual actions can really have an impact. But we need to resist that response, because this crisis will get resolved only if we as individuals take responsibility for it.[4]

From insulating our homes to composting, and from purchasing carbon offsets to choosing reusable tote bags, the actions being suggested to us were sincere and they were relevant. Yet taken in isolation, they were also radically out of step with the scale of the crisis that Gore had so urgently outlined. Sure, Gore did urge us to become politically active, and there was a short section on "becoming a catalyst for change," yet I couldn't help but feel that this call was overshadowed by the prominent encouragement to go to ClimateCrisis.net and calculate your own personal carbon footprint.

In other words, the most prominent and widespread call to arms on the climate crisis was proposing that you go online and calculate exactly how much of this was your specific fault. This was the very same action that you'll remember BP—one of the companies which got unimaginably rich by creating the crisis in the first place—had originally popularized during its deeply deceptive Beyond Petroleum rebranding efforts.

The Rise of Eco-Individualism

Of course, Al Gore and BP weren't the only ones pointing to individual action as the primary solution. As mainstream interest in sustainability grew in the years after *An Inconvenient Truth* was released, the media began to pay attention to the topic. One story, in particular, stayed with me. Written by Penelope Green and published in the *New York Times*, the headline boldly declared "The Year Without Toilet Paper." The subject of the piece appeared to be a quirky personal project taking place in New York City:

> Colin Beavan, 43, a writer of historical nonfiction, and Michelle Conlin, 39, a senior writer at Business Week, are four months into a yearlong lifestyle experiment they call "No Impact." Its rules are evolving, as Mr. Beavan will tell you, but to date include eating only food (organically) grown within a 250-mile radius of Manhattan; (mostly) no shopping for anything except said food; producing no trash (except compost, see above); using no paper; and, most intriguingly, using no carbon-fueled transportation.[5]

Under the more gendered title of *No Impact Man*, the project eventually became a book and a popular documentary movie. In doing so, it took the idea that individual action was our greatest point of power to its ultimate extreme. Beavan, a self-described "guilty liberal"—together with Conlin, and their then two-year-old daughter Isabella—battled their way through a fossil-fueled New York City.

They eschewed taxis and takeout, trains and, yes, even toilet paper, all in the quest to learn exactly what it would take to survive without harming the world around them. At one point, Beavan even shut off the electricity to their small Manhattan apartment while Conlin looked on incredulously.

Soon, it felt like *No Impact Man* was everywhere, including media appearances on *Good Morning America* and *The Colbert Report*. I remember being decidedly irritated by the direction that much of the coverage took. However well intentioned, it felt to me at the time that projects like *No Impact Man* gave those most responsible for causing this crisis a convenient out. How could we possibly hold fossil fuel companies to account when—as Mr. Beavan had so helpfully yet laboriously demonstrated—nobody was forcing us to use their products in the first place?

To be fair, as I've since learned by actually reading his book, Beavan himself was clearly irritated by the coverage too. For one, he became profoundly annoyed by reporters' obsessions with his family's personal hygiene choices. In a particularly candid excerpt, he describes his disappointment when the aforementioned *New York Times* profile was originally published:

> I ride my bike home, thinking over the part of the Times story that really disappoints me. The [Year Without Toilet Paper] headline. I feel that it has trivialized my work. It worries me that I've single-handedly managed to make a mockery of the entire environmental movement.[6]

It's also worth noting that Beavan actually spent a good portion of his book actively discussing exactly the conundrum we are now focused on—namely the relative value of living green in a world that's designed for anything but. Here, he recounts what a longtime activist friend said to him about the potential drawbacks of the *No Impact* experiment:

> Of course the corporate media love you. You're out there telling us all that individually we should use less electricity

> and distracting everyone from the fact that industry is kill-
> ing us. You're out there worrying us about littering while
> they get away with killing the world.[7]

Contrary to most of the stories that were written about the project,
it turns out that *No Impact Man* had a far more systemic lens than
my memory of the project would have suggested. Yet knowing
what we know now about the modern media landscape and Western culture's persistently individualistic and consumerist lens, it's
hardly surprising that journalists and talk show hosts were more
interested in retelling a peculiar story about one family's toilet
habits than they were in examining the structural underpinnings
of what made that journey so damn hard in the first place.

Beavan and Conlin were by no means the only family forging
a different path. I had recently started writing for an eco-minded
design and lifestyle site that was only slightly self-deprecatingly
called *Treehugger*. During that time, I came across countless individuals who were intent on transforming their lives into a teaching
opportunity for the rest of us. Whether it was throwing zero trash
into the landfill, or attempting to live without plastic, it felt like
every other week was bringing a new and enthusiastic protagonist
with a fresh take on the eco-lifestyle experiment.

The Real Value of Lifestyle Activism

Right around the same time that Colin Beavan and Michelle Conlin
were setting out on their journey toward *No Impact*, Irish author
and activist Mark Boyle was launching his own form of extreme
green lifestyle experiment in England. Much like *No Impact Man*,
Boyle sought to use his own personal choices to tell a bigger story
about what he saw as the waste, exploitation, and overconsumption at the heart of our consumer economy. Calling himself *The
Moneyless Man*, Boyle's goal was to literally live without using any
form of monetary currency for at least a year. (Mark donated all
proceeds from his book[8] toward creating a community space where
people could explore money-free living.) Getting by through a mix

of donations, trade, scavenging, and survival skills, he attracted more than his fair share of critics.

Detractors pointed out that Mark's moneyless lifestyle was only made possible by the very system that he claimed to be challenging. Whether through barter or charity, dumpster diving or hitchhiking, somebody, somewhere, had to be spending money so that Mark or other freegans could live off their waste.

From my perspective, however, that was precisely the point. When Mark took journalists dumpster diving, I'm not sure that the goal was to convince every reader to start eating expired food. After all, if you pull out a perfectly good bag of carrots from a dumpster behind your local megamart, and you then proceed to eat those carrots on camera, the lesson is not really that everyone else should eat dumpster-dived carrots.

It's that those carrots shouldn't have been there in the first place.

Similarly, when Colin Beavan and Michelle Conlin shut off the power to their apartment as part of their *No Impact* experiment, they were by no means arguing that the whole of Manhattan should go without electricity and switch to beeswax candles. Rather, they were demonstrating the absurd lengths that individuals have to go to if they want to even come close to living a low-carbon life. In doing so, their efforts served to highlight the need for systems-level interventions to rectify the situation, as Beavan himself pointed out:

> Here was an area where individual action could not help me at all. If I wanted renewable electricity, I needed power companies to provide it and—at least while fossil fuels stayed cheaper than renewable energy—the government regulation that forced the power companies to do so. That, in itself, made for a worthwhile discovery. It was interesting to begin bumping up against the limits of individual action, to see that collective action was also completely necessary.[9]

In other words, the lifestyle experiment approach to environmentalism doesn't necessarily have to scale in order to make a difference. It just has to teach us something about where our systems fall short.

Exposing the Challenges

On January 1, 2020, my former editor at *Treehugger*, Lloyd Alter, made a bold commitment of his own: he was going to do his best to live a year emitting no more than 2.5 tons of emissions. (A monumental reduction from 28 tons the year before.) That's a goal that is roughly in keeping with the aim of limiting average planetary warming to within 1.5 degrees. At first glance, it was a move that surprised me. After all—as we'll hear later in this book—he knows all too well that corporations are desperate to foist responsibility onto the rest of us. In fact, he once caused quite a stir in green circles by claiming that recycling is "bullshit," a scam to distract us from the fact that products are designed to be trashed.

Yet here he was, proclaiming that we can't simply point the finger at the McDonald's Corporation anymore, and suggesting that the best way to challenge the system is to demonstrate, or at least explore, noncompliance:

> We really have no choice. As we have noted many times recently, we have to cut our carbon footprint in half if we have a hope of keeping global heating below 1.5 degrees. And we don't have until 2030; we have to start reducing our emissions right now.[10]

In embarking on his experiment, Lloyd, like others I have talked to in writing this book, actually found that much of this new commitment was no real sacrifice at all: He finally ditched the car and started riding his bicycle everywhere, an activity he greatly enjoys. He stopped eating red meat, which wasn't a big part of his diet anyway. And most consequentially, he cut back drastically on travel, meaning he now had more time at home. In other words, Lloyd found value and personal benefit in rethinking elements of the fossil-fueled rat race. (He compiled the lessons from this experience into a book, *Living the 1.5 Degree Lifestyle*, due out in the fall of 2021.)

This was no old-fashioned appeal to save the world one bike ride at a time, however. Lloyd knows we need fundamental, structural changes if we are ever going to tackle the emergency we are in. Chatting over the phone one afternoon, Lloyd told me that while he was seeking to inspire people to make changes in their own lives, he had also come to realize that perhaps the single biggest value of his entire experiment lay in exposing how context is everything when it comes to lower-carbon living:

> What I've learned is that if you're a rich boomer who bought a house in a nice part of town and you're close enough to bicycle anywhere and you work from home, then it's actually pretty easy to get close to this type of lifestyle. As soon as you're in a position that you have to commute by car, then all that goes out the window. It was easy for me to change because I had the big things already built in. If someone who lived five miles north of me wanted to do this, they have to make much tougher choices.

As I listened to Lloyd talk, I became increasingly convinced that the either/or debates among climate activists about individual action versus systemic efforts had largely missed the point. While once upon a time we had swapped the radicalism of Earth Day for a focus on individualistic behavior change, the idea that we had to choose between the two might finally be receding.

There is now a whole new generation of activists who seem much more at home with the understanding that we're never going to voluntarily slash our carbon footprint to zero through individual actions. Yet still, a fair few of those young activists are also going to impressive extremes in order to forego fossil fuels themselves.

Let's meet some of them.

4

Enough Already

In March of 2020, *Sierra Magazine* published an article looking back on the 50 years that had passed since the original Earth Day. As part of that celebration, the magazine asked what the future of Earth Day might look like in the midst of an accelerating planetary emergency. Among the many prominent voices featured was Jamie Margolin, a Colombian-American teen activist who—along with her friends Nadia Nazar, Madelaine Tew, and Zanagee Artis—had co-founded a youth climate justice organization called Zero Hour.

Jamie had become increasingly politicized in the aftermath of the 2016 presidential election. From helping to organize a Youth Climate March with a presence in 25 different cities around the world, to signing up as a plaintiff in a lawsuit against the state of Washington for failing to safeguard her future, the focus of Jamie's work—like that of many of her peers—has always been explicitly political, systemic, and intersectional in its nature. As such, she had some pretty scathing things to say to *Sierra Magazine* about the watered-down version of Earth Day that she had come to know throughout her childhood:

> I would scroll through articles about how we are living through a mass extinction and then turn on the local news to see the anchors doing fun Earth Day arts and crafts with recycled materials instead of talking about the existential

crisis we are facing and how it affects local communities. It was as if one day a year everyone said, "Let's do the absolute bare minimum for the planet," and then went about business as usual, feeling good about themselves, when I knew that what needed to happen to save my future was revolutionary, systemic change.[1]

This description struck home for me. Thinking back to my own youth, I remembered a publication called *Positive News* that used to circulate widely at festivals and environmental protests in Britain. While the rationale had always been well meaning—namely to counter the perceived negativity of the mainstream press—it always felt like a tone-deaf exercise. Even as the broadsheets at newsstands warned of coming wars, famine, social inequity, or the hole in the ozone layer, *Positive News* would counter with some version of "Class of School Children Recycles Coke Cans."

"*Positive News* makes me want to cry," I remember remarking to an equally cynical activist friend.

Having been connected by a mutual acquaintance in climate circles, I reached out to Jamie Margolin to discuss why she had become so disillusioned with Earth Day. Speaking to me via webcam, she quickly zeroed in on exactly what it was that made her react so viscerally:

A lot of my negative reaction to Earth Day came from all the greenwashing I would see.... Companies would post some generic picture of the Earth, and they'd write something like "Happy Earth Day" or "Let's take care of our common home," and then they'd go straight back to drilling oil or whatever it was they were doing. It just puts a sour taste in your mouth.

I was struck by Jamie's clear-eyed rejection of the kind of "shared responsibility" message that has too often shaped the discussion around the climate emergency for as long as I can remember. As far as Jamie was concerned, it was obvious that there were specific villains in this story. It was also apparent to her that our role was not to help those villains share the burden. Instead, it was to hold them accountable and build a system in which they no longer wielded so much influence over our lives.

As we continued our discussion, I was curious if Jamie had a vision for what Earth Day might actually look like if it were reimagined and revamped for the times we are currently living through. Again, she did not miss a beat:

> It would be a lot more about protesting and demanding change. And it would be a lot less comfortable. Right now, it's just this happy-go-lucky thing where folks are like, "Look at this picture of a flower, let's protect the Earth," as opposed to "These are the corporations that are destroying everything, let's all take action and hold them accountable."

When we spoke, Jamie had recently published a book for fellow youth activists called *Youth to Power*. I was in the midst of reading it at home, and I couldn't help wishing that it had been written when I was in my teens. I remarked to Jamie that it was interesting to me that she had not written one single word about how to lower your own carbon footprint, or how to "live greener" within the confines of our current society. If I read her facial expression correctly, she was perhaps a little puzzled as to why I would even ask such a question. While she told me she had no interest in discouraging individual lifestyle change, she was unequivocal about where we should be focusing the lion's share of our attention:

> I find it nonsensical to be pressuring little kids to be recycling more and then not pressuring massive corporations to stop burning the Amazon. My priority is always holding the powerful accountable. We are not in this because a kid used a straw once. We're in this because massive corporations are profiting off the destruction of life on Earth.

Jamie is by no means alone in her righteous anger.

The Emergence of a Movement

Around the same time that Zero Hour was picking up speed in the United States, Swedish teenager and climate activist Greta Thunberg began her school strike, sitting alone outside of the parliament building in Stockholm every Friday. Here was a fierce young schoolgirl standing up against some of the most powerful people in her country. Similar examples were cropping up in almost every corner of a rapidly overheating globe.

Whether it was Indigenous youth like Jasilyn Charger resisting the Dakota Access Pipeline, or young activists rappelling from a bridge in Portland to prevent a Shell oil support vessel from heading out to the Arctic, a generation was clearly realizing that the adults in charge had no intention of saving them. These young people were willing to put their educations, their reputations, and sometimes even their lives on the line in an effort to secure a future.

In his must-read book *Future Earth*, meteorologist and climate writer Eric Holthaus details the work of Selina Leem, a young climate activist from the Marshall Islands whose address to delegates at the Paris climate summit in 2015 received a standing ovation. Some even credited it with salvaging the summit. Delivering fiery wisdom, Leem refused to accept what so many have already taken as inevitable—that her homeland will sink beneath the waves due to the actions of countries and corporations that have decided these lands and their people are expendable.

As Holthaus explains, Leem is a perfect example of how the sheer urgency of the moment has transformed how people, especially the younger generation, are engaging with the crisis:

> Because the weather is now political, it has generated a social movement. Instead of getting lost in the horror of existential change seemingly beyond their control, people like Selina have helped transform the Marshall Islands, along with other countries on the front lines of the climate emergency, into a place of courage.[2]

In September of 2019, millions of children streamed out of school and took to the streets in order to make their voices heard as part of a global strike. When the events were over, 4,500 different events were recorded in 150 different countries. In Germany, 1.4 million school strikers were reported on one day alone. I joined these strikes in Raleigh, North Carolina, with my then seven-year-old daughter, Adeline. Her older sister, Lilia, joined a march with her mom and classmates down the road in Durham. Listening to speakers on the lawn near the state capitol, and checking out videos on social media from all around the world, I was profoundly moved by the depth and breadth of the movement that these youth-led efforts had been able to convene.

From young, fashionably dressed teenagers to aging boomers and Raging Grannies, and from clergy to young professionals to educators, the crowd was as diverse as it was loud and energetic. Some arrived by foot or bus. Some biked many miles to get there. Others showed up alone in their SUVs. But at least they showed up.

This was, after all, a gathering of individuals who shared one central conviction: we must act now to slow the accelerating climate crisis before it's too late. There's no way such a coalition could have come together if there was a purity test as the basic cost of entry. By identifying both culprits and solutions, and by practicing clear, disciplined messaging, it finally felt like a movement was emerging on a scale commensurate to the issue we were seeking to address.

Identifying the Culprits

That messaging discipline was on full display in Davos, Switzerland, when Swedish school striker Greta Thunberg was asked to address some of the most powerful people on the planet. In a now famous speech to the world's financial elite (and Bono), she directly addressed the main problem with platitudes about shared responsibility:

> If everyone is guilty, then no one is to blame, and someone *is*
> to blame.... Some people, some companies, some decision
> makers in particular know exactly what priceless values
> they have been sacrificing to continue making unimagin-
> able amounts of money, and I think many of you here today
> belong to that group of people.[3]

If the news reports at the time are anything to go by, the audience
in the room was profoundly shaken by this brave act of truth telling. It was a feeling that they might need to start getting used to.

The Rebels Are Angry

In April of 2019, the streets of London largely ground to a halt.
From Parliament Square to Oxford Circus, groups of activists
began gluing themselves to oil company offices, staging die-ins in
the street and, perhaps most memorably, locking themselves to a
giant pink boat in the middle of one of the busiest intersections in
the city. They held some of these locations for nearly a week.

One particular protester captured the attention of news media,
and for very good reason. Farhana Yamin, a leading environmental lawyer who had played a pivotal role in negotiating the Paris
climate agreement, stepped out of the crowd and superglued herself to the pavement in front of Royal Dutch Shell's headquarters.
Asked why a respectable lawyer such as herself would take such a
subversive action, Yamin was succinct in her reasoning: "Writing
books wasn't working."[4]

This outbreak of direct action was the latest salvo from Extinction Rebellion, a loosely aligned coalition of environmental groups
and protesters who were demanding that the government declare
a climate emergency. Inevitably, in a group like this, there were a
wide range of competing messages and demands, including plenty
of vocal advocates for veganism and low-carbon living. Publicly,
however, the early focus was squarely on the government. And it
appears that this ability to punch up, rather than punch outwards,

has inspired a much larger contingent of would-be Rebels, including some who had previously become disillusioned.

Who Is Holding Us Back?

Zoe Young is a UK-based film maker and activist who cut her teeth in the anti-road protests of the '90s and alter-globalization and peace movements of the early 2000s. Having become disheartened over many years by what she describes as "activism's reproduction of impotence," she was re-inspired by Extinction Rebellion, or XR as it is colloquially known:

> I'd retired from activism. We were being bullied into tactics that were getting us nowhere and it was making me sick and miserable. But the vibe is now shifting — and it's not just the younger generation making that change.

Crucially, says Young, the Rebellion didn't simply bring together the anarchists, radicals, and veteran road protesters one might expect to engage in civil disobedience. While some of those were certainly present and contributed their expertise, the movement was built by what she describes as "communities of science, spirit, professional practice, art, farming, education, and government." In other words, ordinary concerned citizens who might have previously kept to more respectable forms of political engagement than this mass lawbreaking initiative. As with the school strikes, it's hard to imagine such a wide coalition could be built if there was any form of lifestyle litmus test before a person could actively take part.

What made the early days of XR so effective, says Young, was its participants' determination to embrace tactical and ideological differences. A core theme of that emerging unity was identifying where the problems really lie:

> Strategic actions must follow the money: Where is the investment coming from that fuels our addiction to fossil

fuels? What tax and subsidy regimes make our society destructive rather than constructive of a future for the kids?

The Personal Is Political (As Long As You Make It So)

For all the systemic activism, however, the push for personal responsibility has not gone away. And neither XR, nor Thunberg, nor
other activists of their generation, are rejecting the value of living a
lower-carbon lifestyle. Nor are they denying the unavoidable truth
that we will all have to change how we live within the space of a
decade or two.

In fact, as has been documented in interviews and articles
around the world, Greta Thunberg personally encouraged her father to become a vegan, and has been vocal about her own refusal
to fly. This refusal eventually convinced her mother to give up air
travel entirely, a commitment that was made all the more significant by the fact that she is a successful, internationally known
opera singer. (Thunberg has also spoken more broadly about the
fact that those of us enjoying relative material affluence are going
to have get used to the idea of limiting our consumption.)

She further demonstrated her commitment to personal lifestyle
change when she traveled across the Atlantic for a meeting at the
UN in a sailboat. Yet her focus even then was on politics and power
first. While the coverage of *No Impact Man* back in the early 2000s,
for example, seemed to too often begin and end with the idea of
rethinking our own choices, Thunberg explicitly presented her
low-impact exploits—even when they involved personal lifestyle
choices—as one more lever among many to demand systems-level
change. When she stepped off yet another yacht as she returned
from the Americas, Greta was quick to disavow the idea that she
was asking all individuals to make the same sacrifices she had:

> I'm not traveling like this because I want everyone to do so.
> I'm doing this to send a message that it is impossible to live
> sustainably today and that needs to change. It needs to be-
> come much easier.[5]

In other words, Greta was not advocating we save the climate one transatlantic yacht journey at a time. Instead, she was using her yacht journey and her no-fly commitments to demonstrate just how carbon intensive our current societal structures are, and she was then using that demonstration to shame those who have stood in the way of change. In doing so, it's just possible that Greta and her generation have found a new way to also sell low-carbon lifestyle choices to a broader audience, too.

A Latent Force

That's a topic that came up when I talked to Ketan Joshi via webcam from his new home in Oslo. (Ketan and I had connected after lamenting on Twitter about the challenges of being immigrants who would rather not fly.) Ketan is an Australian renewable energy expert and the author of *Windfall: Unlocking a Fossil-Free Future*. During our conversation, he talked about what he described as a latent potential that's building in these youth movements to not just take political action, but to also scale lifestyle change so it actually makes a difference:

> The social contagion effect will work well for these younger climate activists and groups. Part of what they do so brilliantly, and on a scale that we haven't really seen before, is encouraging mass participation in change. Of course, that was related to protest and strikes in 2019. But I think a next important step is taking their amazing skills at encouraging public participation in climate action, and saying, "Well, you protested, and it felt amazing. Now you can get that same feeling the next time you make this change in your lifestyle, or your behaviors. You're not just doing your bit for your carbon footprint, but you're part of this mass mobilization that's substantial and that you should feel good about."

While Ketan was excited about the potential for this younger generation to promote behavior change, he still emphasized the importance of focusing on power differentials, structures, and systems as our ultimate end goals. He also issued a word of caution about avoiding the incessant infighting and purity tests that so often consume our energy and distract our movement:

> There are people with incredible power who could easily make decisions to reduce their influence on systemic issues.... And attacking those people, or shaming them, is a powerful tool. Other people are in positions where they...do things that wouldn't align with what they'd explicitly state as their views on climate. You and I are in that category. 99.99 percent of all climate people fall very much into that category. The hypocrisy argument becomes powerfully destructive when it's used in a clumsy way. The problem is that everybody trying to change the system—which 100 percent relies on fossil fuels—has no option but to live within the thing that we are trying to change.

What Ketan was pointing to, I believe, was the idea that behavior change only matters if and when it becomes a catalyst for societal-level change. As such, we need to learn to talk about it differently, and we need to shift our emphasis to mass, strategic mobilizations on specific points of leverage.

While previous generations might have had the luxury of believing that a slow, collective awakening might eventually shift us to a gentler path, a younger generation is calling bullshit on such naivety.

Not everyone, however, has gotten the message.

5

Guilt Trip

In 2008, my wife Jenni and I bought our first house in rural Orange County, North Carolina. It was a beautiful, three-bedroom home surrounded by woodland. There was a small creek meandering through the woods, wildflowers along the graveled driveway, and an abundance of deer that would munch on whatever plants we tried to get started.

While a full green renovation of the house wasn't really feasible, we did set about doing what we could to improve our ecological impact. I joined a co-op and started brewing my own biodiesel. We erected a cedarwood fence and planted a garden. I built some huge compost bins and started collecting horse shit from my neighbor's property. (Horse shit is surprisingly heavy when lugged by wheelbarrow through a forest.) We also converted the old dog house into a chicken coop, installed a more efficient HVAC unit, and (of course) switched out the incandescent lightbulbs for more energy-efficient options.

Then we started exploring the possibility of going solar.

We soon learned that solar electricity would not be cost effective due to shade trees. But the folks at Southern Energy Management, a certified B Corp (more on what that means later), explained that solar hot water was still a realistic option on our south-facing roof. As a long-time clean tech enthusiast, I was excited about this move. Heating water makes up for about 17 percent of a household's energy use. In the warm climate of North Carolina, we calculated that our new panels would cut that portion of our energy consumption by more than half. So we took the plunge.

Already a believer that individual action matters most in how it influences others, I wrote several articles for *Treehugger* in which I bragged about our new array and shared some lessons from the installation process. While much of the feedback was positive, one particular commenter stood out. The reader had deduced, based on the photograph that accompanied the article, that Jenni was several months pregnant. And they felt it was their responsibility to speak up about this outrage.

How dare I, the argument went, talk about my efforts to go green when it appeared that we were about to add another mouth to the world's population? Didn't I realize that the environmental impact of another human being would dwarf any savings from our new rooftop gizmo?

By this point, I had been writing for *Treehugger* for a couple of years. I was well aware of the bloggers' mantra to "never read the comments." Still, troll or not, this particular comment stung. It's not that I even entertained their viewpoint for a second. It just sucks to make what you consider to be a significant contribution toward ecological sanity, and then to be judged or shamed for not living up to someone else's standards.

I'm not the only person who has experienced such attacks.

Eating Our Own

Meg Ruttan Walker is a climate activist, organizer, and mother based in Waterloo, Ontario. When she is not out helping to co-ordinate climate strikes, Meg is more than willing to share her observations on the movement via Twitter. She has been vocal about the tendency for activists to turn inward and eat their own. Even more specifically, she has been righteously savage in her rejection of what she sees as anti-baby misanthropy.

When I mentioned my experiences with the Baby Shaming Troll over the phone, she clearly shared my pain. Much of her work as an organizer is driven by her desire to maintain a livable planet for her young son, and she has sacrificed a good deal of family time in pursuit of real change. Yet, she says, there is a certain class of online stranger who is more interested in judging her decision to procreate than they are in rolling up their sleeves and helping with the job at hand. She's pretty adamant in her rejection of such perspectives:

> How dare you put that on my son and suggest he shouldn't be in this world? He is the reason I do what I do and I'm not having it. Babies are not the reason things are shit.... You don't get to deny people their babies. It's misanthropy and it's eco-fascism.

Just in case the message hadn't come through clearly enough, Meg continued:

> "Keep my son's name out of your fucking mouth or I will come for you."

Clearly, as Meg's fury shows, the act of judging others for something as fundamentally human as making more humans is not only immoral and largely evidence-free (overconsumption is considerably more problematic than overpopulation)—but it's also strategically absurd. It divides the movement and undermines efforts to build solidarity. It also distracts us from focusing on the root causes of humanity's impact as a whole. Unfortunately for Meg and other activists, children aren't the only thing that will earn you a ticket from the purity police.

Undermining a Hero

Few people can be more familiar with environmental shaming than Professor Katharine Hayhoe. According to her website, Hayhoe is a "climate scientist, a professor in the Department of Political Science, a director of the Climate Center, and an associate in the Public Health program of the Graduate School of Biomedical Sciences at Texas Tech University."[1] Her work has resulted in over 125 peer-reviewed papers, abstracts, and other publications. Through her writing, speaking, and research, she has helped countless individuals recognize the existential threat that climate change poses to our wellbeing and way of life.

Yet that, for many people, is never going to be enough. You see, Katharine sometimes gets on an airplane.

The fact that she occasionally travels by air has led some to dismiss her as a hypocrite. Hayhoe appears to have little patience for it. In August of 2019, she shared some of her own experiences as a target of shaming with CBC radio host Matt Galloway.[2] Recalling a meeting she had attended about climate messaging to faith groups, she described how a person sitting next to her had asserted that when she drove her car, she was committing a sin. She was rightly incensed:

> I remember that vividly because my visceral reaction to somebody saying that to me was, "Oh so when I take my child to the doctor, you're saying I'm sinning? When I go to work to support my family, I'm sinning?" And that shaming made me want to just go out and find a Hummer and drive circles around that person.

To be clear, like so many others featured in this book, Hayhoe is not against individual action or behavior change itself. In fact, she has made considerable efforts to walk the walk. And she has also been vocal about sharing what she does to limit her own impact with others.

In fact, she has run experiments on the relative effectiveness of virtual versus in-person talks. As a result of those experiments, she now conducts 75 percent of her speaking engagements as virtual appearances. She also bundles any flights she does take as much as possible, and she has changed her shopping habits to reduce food waste. She counsels her audiences on how to take steps to green their own lifestyles too. Yet given that she's unlikely to eliminate her own carbon footprint any time soon, it's clear that she would much rather lead by example and encouragement than spend a second judging people who choose differently.

This type of approach has always made intuitive sense to me. For all my own efforts to improve my environmental impact, I'm very aware of just how much further I have to go, and how often I make suboptimal choices myself. I've also never felt comfortable telling others how to live their lives. While I've occasionally found friends confessing their supposed "eco-sins" to me in person— whether it was buying a non-hybrid car, or flying abroad for a vacation—I've usually spent more time suggesting other ways they can have a positive impact, rather than focusing on where they (or I) fall short.

That said, I'm not against the use of shaming to bring about change. In fact, I've been involved in shaming efforts myself. When done right, it has proven to be an incredibly effective tool. But you have to pick your targets carefully.

The Power of Shaming

On the 7th of February, 2020, I drove my small, ugly electric Nissan Leaf to a branch of Enterprise car rentals in Durham, NC, traded it in for a small, ugly gas-powered Ford Fiesta, plugged some co-ordinates into my iPhone, filled up a gas tank for the first time in a while, and set off along I-85 toward the Blue Ridge Mountains of North Carolina. (I have once made this 200+ mile journey in my Leaf, but limited range and sporadic charging options made it a one-time experiment in absurdity.) My destination was the semi-annual board meeting of Dogwood Alliance, a forest protection and environmental justice organization focused on defending the unique ecosystems of the US South, as well as the communities that rely on them.

I had gotten involved with Dogwood some ten years earlier, when they had hired my company to develop branding for a campaign against unsustainable fast food packaging. The target was Yum! Brands, an umbrella company which operates KFC, Taco Bell, Pizza Hut, and WingStreet worldwide. Dogwood's goal was to pressure Yum! Brands to stop buying paper made from ecologically sensitive wetland forests, as well as to commit to supporting sustainable forest management schemes like the Forest Stewardship

Council (FSC). They were also asking the company to increase the recycled paper content in their packaging.

The problem was, however, that fast food giants in general—and Yum! Brands in particular—weren't very focused on corporate sustainability at the time. There wasn't much indication that their consumers were demanding it either. Sure, there were some nice-sounding words on their website about being a good corporate citizen and giving back to the community, but there was little doubt that their primary focus was churning out affordable, mass market fast foods at scale. If that meant cutting some environmental corners to ensure the cheapest price possible, then that was a decision they were only too happy to take.

Fortunately, as Dogwood's communications director Scot Quaranda reminded me over beers via webcam, Dogwood was able to identify one thing that Yum! Brands did actually care about:

> The value of their brand is one of the most important things they have. All we had to do was tarnish their brand enough that they want to talk with us and rethink what they are doing.

That's how we landed on a deceptively simple strategy. We just redrew the famous KFC logo with a chainsaw in Colonel Sanders' hands. We juxtaposed it over depressing, aesthetically ugly imagery of the clear-cut forests in the US South. And then we simply named the campaign Kentucky Fried Forests.

"What do we have to do to turn The Colonel into an angry guy?" Scot recalls asking at the time.

This relatively unsubtle approach proved successful. Pressured by the fact that Dogwood had already persuaded other companies like McDonald's to change their ways—and no doubt concerned that we were damaging their all-important brand reputation in the eyes of certain, influential consumers—Yum! eventually announced that it was acquiescing to most of Dogwood's demands.[3] Yet even this sizable victory was not the end game for Dogwood.

On April 10, 2014, International Paper—the world's largest

paper company—signed an agreement with Dogwood Alliance to advance science-based forestry improvements across the whole of the southeast United States.[4] As Scot explained to me some seven years later, the shaming of KFC—not to mention the softer targeting of companies before them—had played a critical role in making this happen:

> This was always the theory of change with International Paper. Most of the brands we went after didn't initially get the connection, but they fairly quickly moved in our direction. We'd picket Staples and Office Depot and shame them into changing the paper they sold. Then we'd go after Johnson & Johnson and Universal Music Group, and explain to them the impact of their packaging. Gradually, as we picked off each target, those that remained looked more and more like the villains. By the time we got to KFC it was easy. They were just such assholes anyway.

Shaping Cultural Norms

I was thinking about the success of the Kentucky Fried Forests campaign as I drove up to the Dogwood board meeting that February. If I was OK with shaming corporate targets like KFC, and if those tactics were successful, couldn't we also use shame tactically to promote behavior change among individuals? And if we could, why did it feel so icky? Fortunately, I was due to have a conversation that afternoon with someone who could help me figure this out.

As the author of *Is Shame Necessary?: New Uses for an Old Tool*, Jennifer Jacquet knows a thing or two about shaming within the environmental movement. An associate professor in the Department of Environmental Studies at New York University, much of Jennifer's work has centered on whether and if shame can be used to shift us toward a more ecologically balanced society.

When I pulled over to call her from a windy rest area just east of Black Mountain, North Carolina, she told me that it's important to distinguish between guilt (a personal feeling we have when our actions are not aligned with our values), shame (a social emotion we experience when our actions go against social norms), and shaming (the public act of exposing a person or entity for transgressing those norms). While they have developed a negative connotation among many, Jacquet argues that there is actually a strong case to be made for the judicious use of all three. They are, she told me, especially useful in situations where legal or political remedies to a problem are not yet feasible:

> Guilt is the best way to regulate society and individual behavior because it's the cheapest form of punishment. If you think about it from a game theory perspective, punishment is costly. You have to take some sort of risk, or pay for a state apparatus to do punishment. If you can get the individual to regulate their own behavior through what we would call a conscience, and if you can get them to internalize social norms, then that's ideal. But anyone who's a parent knows there are a lot of stages to actually achieving that.

Guilt can be useful in shaping individuals' behaviors, and it can be particularly powerful because it works even without an audience. Indeed, I'm not sure there were many people I spoke to in writing this book who hadn't experienced some form of guilt about their impact on climate and the environment. Guilt, however, also comes with its drawbacks. First, it requires an individual to buy into the values or norms being reinforced. Secondly, as a result, it can be very hard to scale.

Shame and shaming, on the other hand, don't actually rely on the presence of guilt or a conscience. As such, they can be used to shape the behaviors of both individuals and larger entities who have not yet bought into the norms being reinforced. That's why, for example, Dogwood was able to shape the purchasing decisions of one of the largest fast food brands on Earth, even though there is little evidence that the corporate culture at KFC placed a high value on environmental stewardship. And it's why Dogwood was then able to use that victory to transform the forest management practices of KFC's paper supplier, too.

Reputation matters to corporations and has a direct impact on their social license to operate. If the reputational cost of their environmental damage outweighs the perceived benefits in terms of financial savings then, all of a sudden, change becomes possible.

In fact, sometimes all it takes is the threat of shaming in order to cause a rethink.

Preserving a Formidable Tool

According to Jacquet, "the threat of shame may be more effective than the actual experience."[5] Because, she says, shame requires the attention of the audience in order to be effective, and because attention is a finite resource, shaming should be used sparingly and strategically for the maximum possible impact.

That's primarily why, says Jacquet, she sees the shaming of individuals for each supposed environmental transgression as a relatively ineffective use of a precious and limited commodity. When shaming focuses on behaviors that are commonplace, widely accepted, and even encouraged by our peers, then audiences are likely to tune it out. That's true even if the person or entity being shamed knows intellectually that what they are doing is counterproductive. And even when it does change some minds, it does not do so on a scale that really moves the needle.

Meanwhile, there is an opportunity cost in spreading the blame around. In other words, it dilutes the power of shaming elsewhere. Much like antibiotics, these strategies work best when they are held in reserve. As Jacquet argues in her book, we need to be thinking about shame and shaming not simply in terms of what or who is morally deserving of shame. Perhaps more importantly, we need to also be thinking in terms of whether or not we are likely to achieve results:

> It is important to understand how responsibility is distributed. In the case of fossil fuels, demand is diffuse, but supply is concentrated—so it might make sense to expose companies for bad behavior. In the case of shark fin consumption, supply is diffuse (sharks are caught by fishermen around

the world), but demand is concentrated among the Chinese elite, so it might be more sensible to focus on demand.[6]

By thinking through the nature of the specific problem we seek to address, we can deploy shame to its greatest possible effect. If activists are going to use shame and shaming to change the behavior of individuals, for example, then they may want to start by targeting specific, influential subsets of society—frequent fliers, for example, or the ultra-rich. (One report by Oxfam has found the world's wealthiest one percent create more than double the carbon emissions of the poorest half of the world's population.[7]) In other words, even as we grapple with who is responsible, we must also ask ourselves a more tactical pair of questions: Who is most likely to cave? And what will the impact be when they do?

The New Pariahs

On the 21st of January 2020, *The Guardian* newspaper reported that Alex de Waal—the CEO of Australian bus company Greyhound—had sent an email to all of his employees earlier that month. In that email, he informed them that the company had signed a contract to help transport workers to and from the massively controversial Adani coal mine that was soon to be constructed in Queensland.

Coming at a time when Australia was still struggling with one of the worst bushfire seasons ever recorded, de Waal rightly predicted that the decision might put the company in the crosshairs of protesters and journalists. Yet, the announcement said, the company had decided to "be courageous" in taking a contract to service one of the dirtiest and largest fossil fuel projects that the region had ever seen.

Activists were quick to respond. They sent letters and emails, made phone calls, and issued statements that called out Greyhound for its role in the project. Billie Tristram, a prominent 14-year-old school striker and climate activist, told Graham Readfearn of *The Guardian*: "It's like a slap in the face. I won't be travelling with them until they back out."

What made this campaign so interesting was the fact that Greyhound was particularly vulnerable to such efforts. As a bus company, it had already been marketing itself as a low-emissions alternative to cars or planes. It had also been building strategic alliances with key environmental organizations.

Ironically enough, CEO Alex de Waal even served as the chairman of the Citizens of the Great Barrier Reef Foundation—a group which has sought to counteract the devastating impact of fossil fuels and climate change on Australia's Great Barrier Reef.[8] Clearly, providing service to a fossil fuel megaproject like this ran contrary to Greyhound's professed ambitions to be a good corporate citizen. Activists pounced on this inconsistency to devastating effect.

One day after the story broke, the Great Barrier Reef Foundation held an emergency 8 a.m. board meeting. Within minutes of the meeting starting, De Waal resigned from the board, and the Foundation later announced that it was severing ties with Greyhound as a corporate sponsor. The organization simply could no longer justify the relationship, even though it had previously brought in AUS$40,000 in funding and promoted more sustainable travel to the Reef.[9]

The pressure snowballed from there. Within a week of that announcement, Greyhound had succumbed. They announced a reversal of their decision and promised not to service the Adani mine project, beyond their preliminary agreement. Activists, who days earlier had been exposing the company as a pariah, were quick to offer redemption. Varsha Yajman, a spokesperson for School-Strike4Climate, thanked Greyhound via *The Guardian*: "We thank Greyhound for not throwing young people under a bus by continuing to help Adani build their climate-wrecking coalmine."[10]

Peer Pressure for the Win

Greyhound's surrender is a near-perfect example of another aspect of Jennifer Jacquet's theory of shaming: namely that shaming works best when the transgressor is a part of, or shares the values of, the group that's doing the shaming. While activists can and do attempt to shame coal or oil companies themselves, those companies are somewhat immune, simply because the transgression is a core part of what they do.

If you try to shame someone's very existence, they are far more likely to fight back than they are to acquiesce. If you instead shame the companies that fossil fuel producers rely on—whether it's banks or software providers, insurers or transportation companies—then you have more leverage. That's because these companies are, more often than not, actively seeking to position themselves as part of the climate solution.

If you can successfully turn these suppliers, then you can hit a triple whammy: you're undermining the smooth operation and profitability of fossil fuels themselves by limiting their pool of vendors, you're inflicting reputational damage on the fossil fuel company, *and* you're redefining good corporate citizenship for the rest of the economy in the process. It's no longer enough for a company like Greyhound to mitigate its own impact. Pressure campaigns like this one are now forcing them to ultimately pick a side.

But what of our own complicity? After all, most of the folks reading this book are, more than likely, still flying in planes, riding in cars, or at least powering their homes with energy derived from dirty fossil fuels. Don't we share some portion of responsibility too?

Guilt Is Good?

I was thinking about this question as I continued my journey up the mountain. Here I was, burning up oil to get to a meeting about protecting our climate. Even as I spoke, other board members were driving or flying in from as far afield as New York City, Connecticut, San Francisco, and Atlanta. This apparent paradox became a topic of conversation over dinner one night. As we carpooled down to a local, farm-to-table restaurant, I asked a few of my fellow passengers how they felt about personal guilt in the face of a planetary crisis. Lynne Young, an environmental engineer based in Atlanta, Georgia, was pretty quick to dispel the idea:

> I don't feel guilty. I do what I can when I can. I've tried to
> improve my environmental impact. But so much of it is out-
> side of my control as an individual. We also have to focus on
> the system.

Also present was a new board member, the podcaster and essayist
Mary Annaïse Heglar. I referenced Mary's essay "I work in the environmental movement. I don't care if you recycle." Chapter 3, and
her broader writings have been immensely helpful to me in exploring this topic. We sat down during a short break in our board ses-
sions, and I asked her whether the title of her now famous article
really was true.

Did she really not care whether or not people recycle?

> Of course I do. It's just that I care much, much more about
> other things. It's great that you're a good environmentalist,
> that you recycle, or that you've changed your diet. I just don't
> think that should be the cost of entry to the movement.

As we chatted, I asked Mary about the apparent contradiction that
she is both a vocal critic of folks pushing individual action as *the*
solution to the crisis, and yet also takes significant steps to limit
her own impact. She eats a vegan diet, for example. She also lives
car-free. She was quick to point out that the choices she has made
were largely those that suited her anyway:

> I went vegan for health reasons first. And I happen to hate
> cheese. I don't own a car, because I don't know how to drive.
> And I moved to the Bronx because I hated being dependent
> on a car, or having to be driven around. These choices were
> easy for me, but they might not be as easy for others.

I nodded along enthusiastically. Yet even though I agreed with
Mary that pointing the finger was largely counterproductive, I also
noted the irony that I still personally feel guilty on an almost daily
basis about what I eat, how much I drive, and countless other fail-
ures to rise above our consumer-driven culture.

Referencing my earlier conversations with Jennifer Jacquet, I suggested that guilt can be a powerful social regulator, and asked whether these nascent signs of guilt among the eco-activist crowd could be a sign of shifting social norms and, eventually, a positive driver for broader social change. She agreed:

> You have to differentiate between shame and guilt. For me, shame suggests there is something fundamentally wrong with you, as a person. Guilt, on the other hand, is something I feel when I've behaved in a way that doesn't match with my values. Guilt can be useful in guiding us toward better behaviors, but we have to ditch the shame.

(In the interests of semantics, it's important to note that Mary Heglar was using different terminology to Jennifer Jacquet. She clarified that her understanding of shame and guilt comes from the work of Brené Brown, and focuses primarily on the difference between character and behavior. Rather than waste time having two writers arguing over the exact meaning of these words, I decided to note the difference and move on.)

Values Are a Moving Target

As I thought some more about the value of guilt, shame, and shaming, I came to realize that not only was there value in understanding them as separate concepts and tools, but also in seeing how they might work together. Shaming—as Jacquet's work has shown—is best used sparingly, and against powerful adversaries who otherwise might not be open to change. That very act of shaming, however, also sends signals out to other elements of society. Just like the kickback from a gun exerts a force on the shooter, shame also has the potential to force change on those of us actually doing the shaming.

Here's what I mean: As protests chip away at the reputations of Big Oil, Coal, and Gas, they also shape the broader social norms under which every one of us operates. If I'm calling a software company out, for example, for working with Big Oil and helping them to make more money, then I am far less likely to work for a fossil fuel company, or invest in them. And I will likely feel increasingly obliged to steer clear of their products too. I might not be able to free myself from their grip entirely just yet, but the stance I've taken publicly about their operations—and those who align with them—also forces me to think twice about whether I really need to drive to the store.

Steve Westlake knows a thing or two about how such ripple effects do (and don't) work. Having been involved with Greenpeace as a volunteer in his teens, Steve actually took a job with Shell just after he graduated from university. That job was then followed by a stint at a motorcycle magazine, and then an international motorcycle racing team—all positions that involved both promoting fossil fuels and a significant amount of carbon-intensive travel. Now residing in Cardiff, Wales—just over the water from the town I grew up in in England—we connected via webcam to discuss his journey. He is reflective now when he thinks back on how and why a climate-aware individual such as himself might have made those decisions:

> How did I come to work for Shell when I had this connection with environmental issues? Back then, they were just as bad—and perceived just as badly too. One of the reasons, I think, is social influence. I just wasn't getting cues from other people that maybe there were better choices to make, both morally and environmentally.

Eventually, as he learned more about the scope and depth of the climate crisis, Steve's conscience got the better of him. He decided he could no longer justify his high-emissions lifestyle or career, and so he changed it. First, he took a position with a cycling magazine. Eventually, he undertook a Masters, and now a PhD, looking at the ways that social influence can shape environmental behaviors. While he too avoids shaming individuals—except in instances where someone is actively bragging about their emissions-intensive choices—he has come to believe that lower-carbon lifestyle choices can indeed shape social norms and create wider-scale shifts in behaviors.

He himself has given up flying, for example. And as part of his research on social influence, he surveyed individuals who knew someone else who had taken this same commitment. The results were rather impressive. Of those people who had social connections that had given up flying, a full 75 percent reported a change in attitude about the importance of climate action and lower-carbon behaviors. Fifty percent even reported flying less themselves. The numbers were even higher when the person in their network was in some way influential or high profile—for example a climate scientist or a celebrity.[11]

According to Steve, the lesson here is not that we should actively shame or expose individuals for environmental transgressions. But nor is it that we should simply ignore personal behaviors or assume that only systems change matters. Instead, he argues, we should learn to have more nuanced discussions where we accept that none of us are going to be perfect, but that strategic and visible reductions in our consumption—especially for high-emissions individuals or influential figures—can help to convey the urgency of climate action and change general attitudes about what's viewed as appropriate everyday behavior (a.k.a. social norms):

> Guilt and shame are highly motivating, potentially. And this is where I believe that the rather simplistic idea, that we should never engage with that discourse, is wrong. They can be a force for change, both personally and collectively.

The trick, both Steve and I agreed, is to find ways to encourage new social norms and cues, while being careful not to distract people from systems-level interventions, nor to let fossil fuel interests off the hook.

That, however, might be easier said than done.

6

Big Oil Wants to Talk About
Your Carbon Footprint

On April 9, 2019, a prominent CEO published an article celebrating the regenerative power of nature. In doing so, he promised up to $300 million of investment in forest protection and restoration. The goal, he suggested, was to repair some of the damage done by the consumption of fossil fuels. There was just one problem: that CEO was Ben van Beurden—the head of Royal Dutch Shell. And it was all in pursuit of something they called "carbon-neutral driving."

Unsurprisingly, this plan from one of the world's biggest polluters was taken with a generous pinch of pink Himalayan sea salt by those demanding climate action. And Van Beurden's own post reveals why:

> Our initiative to offer carbon-neutral driving will begin with customers in the Netherlands. We will then offer similar choices to customers elsewhere, starting with the UK later this year. Shell will offset all the CO_2 emissions associated with our premium V-Power fuel bought in the Netherlands, at no extra cost to customers. For those customers who buy our regular fuel, we will offer them the option of paying one eurocent a litre to offset their emissions."[1]

In other words, Shell—which was continuing to invest billions in oil and gas exploration in Nigeria, China, Bulgaria, and beyond—was making an offer to a *subset* of customers, in a *subset* of one regional market, to plant or protect *some* trees, and more often than not it was doing so with these customers' *own* money. It was then using that offer to present itself as a good faith player in the fight to prevent climate change.

To be fair, Van Beurden's plan to harness the "power of nature" wasn't completely without substance. And he wasn't simply asking his customers to cough up. In fact, he promised that Shell would pay its share too.

As proof points, Van Beurden touted Shell's investments in electric vehicle charging and renewable energy. He also offered to pitch in to cover the upstream, production-related emissions for those customers who did choose to offset. As such, one could argue Shell was not exactly foisting responsibility, but rather proposing a 50/50 split. For such a split to be actually fair, however, one would need to accept the premise that the blame for the problem is also equally shared.

Unfortunately for Van Beurden, history would suggest otherwise.

Some Are More Responsible Than Others

In 2018, the Dutch news organization *De Correspondent* released internal documents revealing that Shell had hidden its own scientists' warnings about climate change since at least the 1980s. In one particular report, written in 1988 and discovered by journalist Jelmer Mommers, scientists recommended that the oil sector—and society at large—needed to take action immediately, before the worst impacts of climate change started to manifest themselves:

> It is estimated that any climatic change relatable to CO_2 would not be detectable before the end of the century. With the very long time scales involved, it would be tempting for society to wait until then before doing anything. The potential implications for the world are, however, so large that policy options need to be considered much earlier. And the energy industry needs to consider how it should play its part.[2]

And yet despite these very clear warnings from their own scientific teams, Shell would continue for decades to fund lobby groups and public influence campaigns to obscure or deny the science around climate change, while simultaneously putting obstacles in the way of potential legislative solutions.

Given the context of his organization's efforts to actively hide its own scientific findings, Van Beurden's offer on LinkedIn to "share responsibility" with consumers feels a little akin to Pablo Escobar offering to go Dutch on rehab bills with some of his end consumers. Sure, there was a shared responsibility in that a person *could* choose to not use cocaine, but given both the addictiveness of the substance and the nefarious business practices involved in its promotion, it seems fair to place a larger portion of the blame on the production and distribution end of the equation.

But I digress.

Shell was by no means alone in its complicity. In 1983, Exxon scientists predicted that atmospheric concentrations of CO_2 could reach 415 ppm (parts per million) by 2019, and that these concentrations would result in an average temperature increase of around 1 degree Celsius (1.8 degrees Fahrenheit). This prediction proved to be so accurate that Brian Kahn—editor of the Gizmodo publication *Earther*—told me that it gave him chills to reread the memo some 37 years later:

> It really is uncanny that Exxon head scientists were working with much simpler models in the '70s and '80s. And even still, those models showed a striking symmetry in terms of the trajectory of emissions, and the trajectory of temperature that followed it. The mesh, particularly when we crossed that threshold of 415–420 ppm, is uncanny how well it synchs up [with what has actually happened]. And it speaks in a lot of ways to the fact that Exxon scientists knew what everyday scientists knew back then, and instead of acting on it to change their business model, they acted on it to change the public's perception of the science.

By denying climate change for as long as they could, and then opposing, sabotaging, and delaying any meaningful action, fossil fuel companies have sought at every turn to frame debate in the most favorable terms possible for fossil-fueled business as usual. First, that involved outright denial. And, as Michael E. Mann has documented in his 2021 book *The New Climate War*, it later involved emphasizing and promoting "solutions" that would allow the gravy train to keep rolling for as long as it possibly could. When all else failed, they then sought to present themselves as magnanimous conveners of a "conversation" about how we could possibly now find ourselves in such a pickle. It's a page out of a corporate playbook that's been deployed many times before.

The Tobacco Playbook

For years, tobacco companies resisted efforts to curb smoking. But as evidence mounted, they eventually agreed to pay for advertising on the health impacts of tobacco. They did so, however, having secured an agreement that those ads didn't vilify or assign blame to the companies themselves. In other words, they were all for engaging with the issue, as long as that engagement didn't result in a negative impact to the allure of their brand.[3]

Beverage and packaging companies had pulled a similar move following the introduction of disposable bottles in the 1950s. As potential bans were floated in response to the growing plague of trash, the industry came together to promote anti-littering campaigns and then, eventually, recycling. As I mentioned earlier in this book, environmental writer, former colleague, and would-be 1.5 degree lifestyle leader Lloyd Alter has some strong opinions about the moral culpability of industry in this scheme. Here's how he called out such efforts over at *Treehugger*:

> Let's call recycling what it is—a fraud, a sham, a scam perpetrated by big business on the citizens and municipalities of America.... Recycling is simply the transfer of producer responsibility for what they produce to the taxpayer who has to pick it up and take it away.[4]

As we were having drinks via webcam one evening, I suggested to Lloyd that the maneuvers by Big Oil to frame the climate debate as a question of individual responsibility could be seen as a similar effort in distraction. His response was both a yes and a no. On the one hand, he agreed that it's convenient for fossil fuel producers to keep the conversation focused on the individual scale. On the other hand, he argued, those of us who are aware of the climate crisis and have the means to make better choices probably shouldn't be let off the hook either.

A few days after we spoke, a social media storm blew up once more when Shell Oil's social media team decided to ask its Twitter followers what they would be willing to change in order to combat climate change. The response had been predictably derisive, and Lloyd reached out to me for my take. Here's what I told him:

> Two things can be true at once. Shell Oil has no place asking us about our personal carbon footprints, and also we should probably be asking ourselves about our own carbon footprints. Where it gets murky is how much we environmentalists should be focusing on each other—and certainly when it comes to pointing the finger. Because that can quickly derail a movement.

In other words, spending time interrogating our own lifestyle choices is a valid exercise. Yet we need to be careful that in doing so, we don't inadvertently provide corporate polluters with an assist. In much the same way that big employers would rather negotiate with 5,000 workers separately than they would the collective power of a union, fossil fuel companies and their allies would rather isolate climate efforts to the smallest possible unit of action, namely the individual consumer.

This approach, however, is beginning to run out of steam.

They've Never Been the Good Guys

Sometime around the new millennium, I met up with a small group of fellow students at the University of Hull, in North East England. We marched down to our local Shell gas station, where one of the braver ones among us (not me!) snuck behind the sales kiosk and shut off the emergency valve to the gas pumps. They then secured the valve in the closed position using a bicycle U-Lock, thus preventing the owner from restarting the flow of fuel. We then settled in for an afternoon of arguing with irate motorists and discussing the politics of oil with a very polite British policeman. The tactic might seem familiar to anyone following recent protests against gas pipelines. Yet there was one crucial difference: we were not protesting the sale of oil itself. We were seeking to draw attention to Shell's corporate complicity in the arrest and execution of Indigenous Ogoni activists in the Niger Delta.

That was the nature of much of our activism back then. While we still stood up against the same oil giants, the goal was—more often than not—to encourage better corporate citizenship and an end to the most egregious examples of habitat destruction or human rights violations. Even when we advocated a shift to renewables, I suspect most of us couldn't quite envision getting rid of the transgressors entirely. Or at least, when we did, it felt more like an exercise in utopianism than near-term goal setting for the future of civilization.

All that has now changed.

Deflating the Carbon Bubble

In 2014, the Governor of the Bank of England, Mark Carney—a respected pillar of the economic establishment—dropped a bomb on conventional capitalist thinking. Speaking in language that every economist could understand, he stated that most existing fossil fuel reserves were now essentially stranded assets. The laws of physics being non-negotiable, much of the oil, coal, and gas that had already been discovered would now have to stay in the ground if we were to head off the worst impacts of climate change.[5]

Groundbreaking as this statement was, Carney was only confirming what the climate movement had already been warning of for some time: carbon-heavy industries existed in a bubble which relied on a continued growth in demand, and that growth in demand was fundamentally incompatible with the goal of maintaining a livable climate. One way or another, whether through regulatory action, technological paradigm shifts, societal awakening, ecological disaster, or some combination of the above, that carbon bubble would eventually have to burst. When it did, the fallout would be spectacular.

As report after report set out just how much damage had been done by the carbon we had already emitted, activists started calling not for corporate reforms or incremental emissions reductions, but rather for an ambitious and rapid decarbonization of our entire economy. They then set about to get this done through strategies that included defunding and deplatforming fossil fuel giants by protesting their gallery, museum, and university donations; trolling them mercilessly on social media; as well as demanding legislation to phase out fossil fuels within the space of a few decades.

The urgency of these calls grew ever louder when, on the 8th of October, 2018, the UN Intergovernmental Panel on Climate Change (IPCC) released a landmark report that found we had just twelve years to cut global emissions nearly in half, or else the goal of keeping warming to 1.5 degrees Celsius would likely slip permanently out of our grasp. Even for those of us who had been following the climate crisis for decades, and who knew that things were serious, the gravity of what was being communicated came as a profound and deep shock. I know of more than one adult friend within the movement who has described feelings of existential dread and grief as a result of reading this report.

I too had my moments.

Fortunately, as we've already seen, the younger generation was not ready to take this lying down. Not only did activists get more radical in terms of their demands as these stakes became clear, but they also escalated the fight in terms of tactics too. In his book *What We're Fighting for Now Is Each Other*, author and activist Wen

Stephenson documented the birth of this new movement. In doing so, he laid out the profoundly moral terms under which we must now conduct the climate debate:

> ...given what we know and have known for decades about climate change, to deny the science, deceive the public, and willfully obstruct any serious response to the climate catastrophe is to allow entire countries and cultures to disappear. It is to rob people, starting with the poorest and most vulnerable on the planet, of their livelihoods, even their lives—and their children's lives, and their children's children's lives. For profit. And for political power.
>
> There's a word for this: these are crimes. They are crimes against the earth, and they are crimes against humanity.[6]

From valve turners facing trial for shutting down pipelines, to a small group of citizens piloting a lobster boat into the path of a gigantic coal tanker, the tactics that Stephenson depicted in his book were confrontational, disruptive, and quite often illegal—just as they were back in my student days too. The difference was that these protesters were now willing to call out fossil fuel giants, not just for specific and localized crimes, but essentially for their very existence.

So what, in the face of such existential opposition, is a good, ethical oil major supposed to do?

Can Big Oil "Go Green"?

Interestingly, there are early signs that some companies, on the face of it at least, are trying to grapple with the question of what a "sustainable" business model might look like for them now. On December 4, 2019, Repsol—Spain's largest oil and gas company—made a surprising announcement: It was aiming to become a net zero emitter of carbon by 2050 at the latest. Even more surprising was the fact that this commitment included emissions from the end-customers using their products.

To back up their words, the company even promised to tie the bonuses of executives specifically to the push for reduced carbon development. Parts of the proposed plan, such as increased investments in renewables, for example, were not without merit. Others, however—like developing as-yet-unproven-at-scale carbon capture and storage technologies, or using forestry to offset emissions—were more problematic. And despite a promise to prioritize "value generation over production growth," Repsol was quick to point out that it saw a long future for oil and gas, even in a lower-carbon economy.

Whatever the substance of such announcements, the very fact that they are being made is a sign that oil companies are beginning to feel the heat. Unfortunately for them, however, it's almost certainly a case of much too little coming far too late. As Lloyd Alter said to me, the sheer scale and nature of oil companies' investments means that any climate action that's compatible with a livable future will almost certainly result in their demise:

> You are who you are and you are good at what you are good at. Kodak was unrecognizable after the switch to digital photography. And oil companies won't survive the low-carbon transition. At very least, they'll be smaller and very, very different. Sure, if we were still talking about resource efficiency and a gradual transition they might stand a chance. But it's increasingly clear we need a rapid shift and a fundamental break with the past. "Keep it in the ground" is a much different idea than "use what you have wisely."

Time will tell if Lloyd is right. If he is, however, then oil companies will have only themselves to blame.

A Missed Opportunity

Climate journalist and podcaster Amy Westervelt knows more than most about the surprising twists and turns in the story of oil and climate. Having been laid off when the first dot.com bubble burst in the late 1990s, she reluctantly took a job editing technical reports for a consulting firm. On a Zoom call with her during the

2020 coronavirus lockdowns, she described to me how one of her early assignments served as a kind of wake-up call:

> One of the case studies I had to write about was them re-engineering some of Shell's offshore oil platforms for sea level rise. And this was in 2002, or 2003. And I remember then thinking "this is interesting," because I feel like oil companies mostly act like climate change isn't happening. And yet here they are reengineering their oil platforms.

Having eventually escaped from her technical writing gig, Amy spent years as a journalist grappling with how to write about climate change. She became increasingly frustrated with the failure of the movement to break through with its messaging. The challenge, she realized, was that the story was always told in terms of the physics and data, not how this crisis would show up in our daily lives. She also realized that most of the best stories in our culture feature a villain. How could you tell the story of climate change without talking about who was most responsible for how it came to be in the first place?

Fast forward to 2018 and Amy was covering the story of a class action suit against oil companies for their role in contributing to climate change. The judge in the case asked for a tutorial on the topic, and Amy realized that this was her chance to really tell the story in a meaningful way. True crime was exploding as a genre at the time, particularly in the world of podcasts, and Amy decided to launch *Drilled*—a podcast that applied the stylistic and narrative elements of true crime storytelling, but to one of the single biggest corporate crimes in history: how Big Oil had actively sabotaged societal efforts to stop the climate crisis.

Interestingly, *Drilled* didn't just tell the story of Big Oil's moral failings. It also pointed to an alternative history that could so easily have come to pass.

In the very first episode, titled "The Bell Labs of Energy," Amy explored how scientists and executives at Exxon during the mid-seventies had sought to position the company as a leader, both in climate science and the development of solutions. Gradually, however, Reagan-era resistance to regulation and government intervention led to more obstructive strategies prevailing. This cultural shift set the scene for the disinformation and lobbying campaigns that the oil companies ultimately chose to focus on.

Chatting with Amy, we discussed the question of what a robust industry response to climate change might look like in 2020. We couldn't help but both harken back to what might have been:

> That's the thing that kills me about the Exxon story in general. If they had kept on that path, we would absolutely be living in a different world today. The guys at Exxon came up with the first lithium ion batteries. They were really getting places on solar and wind, and even just understanding how climate change works.... You had a group of people who thought the way forward was to understand the science, earn a seat at the table, and help shape the solutions. It went from that to a group of people who thought the best way to deal with regulations that are coming your way is to stop them.

As someone who has talked to countless activists over the years, I am used to hearing anger and suspicion when they talk about fossil fuel corporations. But Amy's voice betrayed entirely different sentiments—something more akin to disappointment and incredulity. We could so easily be enjoying an entirely different future right now, she lamented, one where we were well on the way to fixing climate change and moving on to other challenges.

In sabotaging that future, oil companies haven't just betrayed society as a whole. They've also undermined their own ability to survive.

Balancing on the High Wire

On January 15, 2020, BlackRock CEO Larry Fink sent a letter to the CEOs of all of the corporations that his investment management company invests in. In this latest iteration of "Larry's Letter," which has become a closely watched tradition in financial circles, Fink described the world as being on the edge of a "fundamental reshaping of finance." And he left no doubt about the fact that climate change is the driving force behind that shift. (Before we get too carried away, however, it's worth noting that Fink's company is still one of the world's largest investors in fossil fuels—not to mention a consistent opponent of shareholder resolutions on climate.)[7]

Still, the fact that Larry Fink was even signaling a shift away from fossil fuels was a pretty big deal in the eyes of the financial world. Citing past crises ranging from the inflation spikes of the '70s and '80s to the Great Recession, Fink argued that climate change poses a different, more long-term and structural challenge to the global financial order. Specifically, Fink pointed to investor perception of future risk as a potentially disruptive force:

> ...because capital markets pull future risk forward, we will see changes in capital allocation more quickly than we see changes to the climate itself. In the near future—and sooner than most anticipate—there will be a significant reallocation of capital.[8]

In other words, if an asset manager is considering financing a new mine, a new pipeline, or a new power station, and if they then come to believe that robust carbon pricing or disruptive tech might be coming in 10, 20, or even 30 years' time, this changes the calculation on what they will do *right now*. Once those managers shift their money from industries that represent the problem, and instead place their bets with industries that are working toward solutions, the pathway for us as individuals to live zero-emissions lives becomes that much easier to take.

Coal as the Canary

Writing on Twitter, futurist and author Alex Steffen pointed to the sudden and unpredicted collapse of coal industries in the 2010s as a harbinger of what might come to be. Crucially, he argued, the threat does not just apply to parallel energy interests like oil and gas, but also to a whole host of related industries that ranges from automotive to aviation to architecture and beyond:

> It's worth remembering that coal is the canary in the financial mine. Whole industries, tens of thousands of companies in different sectors, government bonds, infrastructure projects, real estate, etc.—a huge swath of the modern world—is at risk for rapid repricing now.[9]

What's interesting in this regard is that the world didn't have to stop using coal for the writing to be on the wall. Instead, it took a combination of marginal improvements in energy efficiency, drops in the price of renewables, and competing pressure from so-called natural gas to put a small dent in demand growth. And that dent was enough to start shifting the financial investment appetite.

All of a sudden, policy makers and financial institutions began to recognize that coal was never coming back. Mining unions began talking about a just transition away from coal.[10] Countries like the UK accelerated their decarbonization. And investors like Larry Fink kicked the industry while it was down by withdrawing their support. Even the election of Donald J. Trump—for all the rhetoric of ending the war on coal—couldn't reverse the decline. In 2018, the US power sector saw the second highest number of coal plant retirements ever.[11]

All this might seem wonky if you—like me—aren't particularly financially or mathematically literate. Yet it is an important insight into how we, as citizens, can move the needle in our favor. It's less important that each of us eliminate fossil fuels from any and all aspects of our lives, a task which requires herculean effort and is likely to only attract a small hardcore contingent. Instead, what matters is that we start chipping away at the broad viability of fossil fuel industries as the perceived drivers of our economy. Each incremental step we take builds on every other one. If we direct those steps to where they have the biggest impact, we will soon find ourselves hitting tipping points that can bring entire industries to their knees.

It goes without saying that ditching the car, using transit, buying an electric vehicle, or turning out the lights can be useful strategies to achieve this goal. So can going flight-free. But these should be seen as one part of a bigger bag of tricks that includes both individual and institutional-level divestment from fossil fuels; policy-level efforts to curb demand or make supply more expensive or cumbersome; as well as community-level efforts to make alternatives more viable. Here's how Brian Kahn—the *Earther* editor I spoke to earlier about Exxon's shenanigans—described the challenge to me during our conversation:

> We do know that there are things we can all do [to reduce our carbon footprint], and we should not be excused for ignoring those things. But to focus on them as the only thing, or the most important thing, misses the fact that there are much bigger actors out there who have much more to do with [creating the problem].
>
> These companies have made billions and billions of dollars from presenting those lies, for decades, and they still form the backbone of our economy. If you're looking at the problem, it's hard not to think, "What can I do?" But we really need to be asking ourselves, "What can I do to hold these companies accountable? And how do we transform our government's relationship with them so that the government is working for the people, not just these business interests?"

Whether it's these companies' reputation or their political influence, their financial performance or the legal environment in which they operate, there are many levers we can pull to start putting obstacles in the path of their hegemony. None of us can do everything. What matters is that we grab a lever (or three) and start pulling.

It's not going to be easy, however. We're going to have to pull those levers hard.

A Tenacious Grip on Power

Back when I talked to Amy Westervelt during the COVID-19 lockdowns of 2020, it was a very strange time. Global oil demand had temporarily fallen some 30 percent as commuters stayed home and planes stayed grounded. Gasoline was selling for around $1.80 a gallon here in North Carolina, and the price of crude oil had briefly gone negative. (That meant that oil drillers were actually paying people to take their oil away because they had nowhere left to store it.) I mentioned to Amy that this felt like a potential turning point for the dominance of oil—and a sign of how precarious that dominance was—and she surprised me with a note of caution. The oil industry, she said, had been in trouble for years. This crisis might actually have bought them a lifeline:

> The entire industry's business model is built around $100 barrels of oil, which is a price we haven't seen since 2016 and are unlikely to see again. And projected demand for transportation—which is how the oil industry figures out how it's doing—had started declining in 2018.... That's before you even consider that the whole fracking and tar sands industries were a house of cards to begin with. They've never once made a profit. Not once. What that tells you is that the oil and gas industry has been propped up by governments. And the reason they've been able to do that is the influence they've been able to create for themselves.

Sure enough, the day after I spoke with Amy I saw an online video of President Trump meeting with Senator Dan Sullivan of Alaska as they discussed how banks were "discriminating" against energy companies by not financing Arctic drilling. This discrimination, said Trump, might be reason enough for the federal government to step in and intervene. In that same week, the *Wall Street Journal* reported on potential government stimulus efforts aimed at keeping oil and gas companies afloat.[12] The old adage, that you should never let a crisis go to waste, had certainly not been lost on fossil fuel companies and their political allies during COVID-19.

Clearly, if we are going to stand any chance of defeating such influential and well-connected opponents, then we are going to need some pretty powerful allies. It's possible that some of them will come from slightly unexpected quarters.

7

Corporate "Citizenship" Reimagined

On November 25, 2011—also known as Black Friday—outdoor clothing company Patagonia ran an ad in *The New York Times* with a rather surprising headline: "Do Not Buy This Jacket"

The core substance of the ad promoted the company's Common Threads Initiative, which was aimed at increasing repair and recycling of Patagonia's own clothes. There was, however, a much larger, more societal-level message. An accompanying blog post laid out the rationale:

> The most challenging, and important, element of the Common Threads Initiative is this: to lighten our environmental footprint, everyone needs to consume less. Businesses need to make fewer things but of higher quality. Customers need to think twice before they buy.
>
> Why? Everything we make takes something from the planet we can't give back. Each piece of Patagonia clothing, whether or not it's organic or uses recycled materials, emits several times its weight in greenhouse gases, generates at least another half garment's worth of scrap, and draws down copious amounts of freshwater now growing scarce everywhere on the planet.[1]

Of course, Patagonia had long positioned itself as an antidote to fast fashion. A cynic could easily (and somewhat rightly) argue that, of all companies, they had little to lose, and a lot to gain, from promoting an economy where consumers value expensive, durable goods over inexpensive, more replaceable ones. So what made this different from just another exercise in corporate virtue signaling, the kind of thing that teen activist Jamie Margolin had so roundly condemned when I talked to her about Earth Day?

"Responsible" Versus "Sustainable"

In 2012, company founder Yvon Chouinard expanded on the thinking behind the Common Threads Initiative in a book he co-authored with Vincent Stanley called *The Responsible Company*, in which they explored how to act responsibly within a system that encouraged the opposite:

> We can't pose Patagonia as the model of a responsible com-
> pany. We don't do everything a responsible company can do,
> nor does anyone else we know. But we can illustrate how any
> group of people going about their business can come to re-
> alize their environmental and social responsibilities, then
> begin to act on them; how their realization is progressive:
> actions build on one another.[2]

Patagonia could and did take steps to reduce the impact of its own products, by using recycled materials, for example, or buying renewable energy. Yet choosing to focus solely or even predomi-nantly on its own environmental impact—even if 100 percent suc-cessful—would only leave it operating as a model for sustainability while those around it continued to drive our climate and ecology into the ground. And that's hardly a recipe for long-term success.

There are obvious parallels here with the question of individual versus collective action.

That's because Patagonia's approach was firmly rooted in the idea of their business as one part of a much broader community. As part of its efforts, the company started conversations and built partnerships with companies including Ben & Jerry's, The Body Shop, Smith & Hawken, REI, and The North Face. It published details of its own supply chains. And it then used this transparency to push similar levels of traceability elsewhere.

Crucially, Patagonia also began to make its political voice heard too.

Corporate Citizenship — For Real

When President Donald Trump and congressional Republicans enacted corporate tax cuts in 2017, most businesses were all too happy to quietly pass the additional profits on to their shareholders. Patagonia, however, did not. In a post written on LinkedIn, Rose Marcario — the company's CEO at the time — announced that they would take the money they had saved in taxes and donate the entire amount to organizations working on climate and the environment. The reasons behind the move were simple:

> Based on last year's irresponsible tax cut, Patagonia will owe less in taxes this year — $10 million less, in fact. Instead of putting the money back into our business, we're responding by putting $10 million back into the planet. Our home planet needs it more than we do.
>
> Our home planet is facing its greatest crisis because of human-caused climate disruption. All the extra heat we've trapped in the Earth's atmosphere is not only melting the poles and raising sea levels, it's intensifying drought and accelerating the extinction of species. The most recent Climate Assessment report puts it in stark terms: the U.S. economy could lose hundreds of billions of dollars, and the climate crisis is already affecting *all* of us. Mega-fires. Toxic algae blooms. Deadly heat waves and deadly hurricanes. Far too many have suffered the consequences of global warming in recent months, and the political response has so far been woefully inadequate — and the denial is just evil.

Clearly, this was not your average sustainability statement. Previously, perhaps influenced by the consumerist and individualist lens of our culture, corporate responses to climate change had tended to focus on being less bad. Consequently, the term "corporate citizenship" was bandied about to describe everything from buying green energy to developing Zero Waste initiatives. Yet it always felt like somewhat misleading terminology.

Reducing your energy use, after all, is primarily about improving your bottom line. In that sense, it's about as close to true civic engagement as remembering to turn out the lights. Patagonia's version of corporate citizenship, however, was more believable. By looking beyond its own worthy efforts at sustainability, and instead asking what it could do to change the structures within which it was operating, it was starting to behave just a little bit like an actual, real-life citizen.

All of us—as consumers, as citizens, as employees, as investors, and as business owners—might want to start paying more attention to how the companies we interact with are behaving in the civic realm.

Luckily for me, that's somewhat baked in to the culture with my own current employer.

A Different Kind of Insurance

If you drive up to the parking lot of The Redwoods Group in Morrisville, North Carolina, you'll see a nondescript corporate office building that could be anywhere in the United States of America. Yet walk into those offices and you'll be greeted by a wide-open floor plan, a friendly café space, and walls emblazoned with the words "Love," "Serve," and "Transform" in gigantic lettering.

This does not feel like the HQ of your typical commercial insurance company.

Redwoods—which I have worked with first as an independent contractor, now as brand development manager—was created by husband and wife team Kevin and Jennifer Trapani in 1998. The name itself stems from the fact that redwood trees are connected through one root system, sharing resources and giving each other strength against whatever storms might come their way.

The company's business model is based on insuring YMCAs,

Boys & Girls Clubs, resident summer camps and Jewish Community Centers. It then uses the insuring relationship to develop tools and strategies to keep kids safe from sexual abuse, drowning deaths, bullying and other forms of social harm. By gathering insurance data and insights from the claims it receives, Redwoods is able to use that data to inform customers about ways they can better protect their community.

When I drove to Chapel Hill, North Carolina, to meet Kevin and Jennifer for socially distanced backyard cocktails in the late fall of 2020, systemic change was at the very forefront of everyone's minds. The presidential election had occurred just two days before, and Donald Trump was refusing to concede. In fact, he was giving a press conference from the White House to talk about "election fraud," "suppression polls," and #FakeNews just as I pulled into their driveway.

Following a brief conversation in which we vented our disgust at the behavior of the outgoing president, we muted cable news and retreated to the screened-in porch to discuss the role of business in creating social change. Kevin wasted no time in distinguishing between run-of-the-mill lobbying, and what he saw as more civic-minded engagement:

> It's not new for businesses to be involved in politics. Businesses are advocating all the time and spend hundreds of millions of dollars lobbying. But they are doing it for themselves. What they need to do is find a way to do it for the good of all—rather than just for the good of self. When they do that, they stand out, and they wield a huge amount of influence. There is power in being the unexpected voice.

Recently, Redwoods has been finding ways to do just that.

Beyond Corporate Responsibility

As the #MeToo movement against sexual assault and misconduct picked up steam, juries began returning verdicts and settlements in cases of abuse that were an order of magnitude higher than historical precedents would have suggested. Legislators around the US were also responding to the changing times. As state after state enacted a wave of reforms to statutes of limitation, they raised the prospect of survivors seeking justice for events that had happened many decades before.

Given what we know now about the prevalence of child sexual abuse in our communities, youth-serving community organizations—and by extension their insurance carriers—were now facing the potential of what Kevin described as a "tsunami of historical allegations." This was an existential threat which could eventually make such organizations essentially uninsurable.

Redwoods was faced with a choice.

On the one hand, the company could actively oppose reforms due to the impact they would have on the broader social mission of its customers. Or, alternatively, it could stand with its values and advocate for robust, survivor-focused reforms, and rally its customers to supporting those reforms too. According to Jennifer—who serves as the company's chief underwriting officer—this really was no choice at all:

> When we were talking about it, we felt we had no choice but to advocate for truth and go in the opposite direction of harm. I wish more business leaders were willing to think this way. I know people who are both very successful professionally, and very moral in their personal lives, but they see business as business and personal as personal. That's dangerous and it's harmful.

There are lessons here for the broader good-for-the-world business community. Just as individuals need to understand and engage with the societal structures they interact with, so too do business leaders. And they need to do so with a firm sense of the common good. After all, it's also in the long-term interests of the business community to make sure that our society functions as it should.

Fortunately, this realization may finally be catching on.

A Different Type of Shareholder Primacy?

The Shareholder Commons is a nonprofit organization which, according to its website, seeks "a shift from unbridled profit-seeking to a capitalism that values only those profits that are good for the planet and its inhabitants."[3] Founded by former corporate lawyer Rick Alexander, the organization's focus is on one very specific part of corporate governance and behavior—namely unlocking the potential of institutional investors with diversified portfolios to moderate the behavior of business as a whole.

For someone like me, who has grown up skeptical of capitalism in general, and shareholder influence in particular, this is a somewhat counterintuitive approach. Rick, however, walked me through the logic when we spoke via webcam. Drawing parallels with the challenges that individuals face when balancing our immediate consumer desires against our more altruistic societal values, Rick argued that voluntary action by businesses will only ever get us so far:

> It is important that companies act in a way that collectively addresses the issue, whether it's climate, or human rights, or inequality. But it's hard for companies within a capitalist system—just like individuals—to actually do that. There's nothing wrong with a person flying less, or a company reducing its energy use. The problem is that in a capitalist system, it doesn't really work, because each individual company is incentivized to maximize profit. It's just how capitalism works.

Many within the movement are working to enact top down change in the form of government regulation. Rick acknowledged that work as extremely important, but then he qualified that The Shareholder Commons is attempting to take on a slightly different piece of the puzzle:

> Investors are generally diversified—so...they don't actually want companies to make a profit at the expense of...the

environmental and social and economic systems that their portfolio of companies is embedded in.

If Rick's thesis was correct, empowering large shareholders to value not just immediate financial returns from each individual company but contributions to societal and ecological health too, could be one important step toward minimizing the damage that capitalism can cause. Of course, there are some companies that are already a step ahead in this regard.

And these companies are beginning to get a lot more vocal too.

Benefit Corporations Step Up

One afternoon in late 2019, Bart Houlahan stopped by the Redwoods offices. One of the original co-founders of famed athletic shoe company AND1, Bart and his business partners had always prided themselves on running a socially responsible enterprise. When their company sold, however, they saw just how quickly a corporation's values can be stripped away.

That experience led Bart, along with his fellow AND1 co-founders Jay Coen Gilbert and Andrew Kassoy, to start the movement of Certified B Corporations (the "B" stands for "Benefit")—of which Redwoods was an early champion. Member companies are assessed regularly on a rigorous set of criteria that covers everything from their waste management practices and energy consumption, to executive-to-worker pay ratios and equal opportunity employment.

On his visit to our offices, however, Bart shared that the rapidly growing movement had been undertaking some soul searching. In the face of the escalating twin crises of the climate emergency and social inequity and systemic racism, he suggested it was time to start thinking beyond simply certifying responsible behaviors. In other words, just as individual citizens needed to think about more than just their personal carbon footprints and their role as consumers, socially responsible businesses needed to leverage their platform and wield their influence for broader, systems-level change.

The Power of Corporate Activism

Unsurprisingly, many Certified B Corps had been speaking out for years already. Ben & Jerry's, for example, had pushed campaigns on issues ranging from the protection of coral reefs to the development of renewable energy. And we've already seen how Patagonia—another certified B Corp—was willing to call out the Trump administration over climate change. What Houlahan was now suggesting was that we should encourage, and perhaps even expect, all B Corps to do the same, and to do so with one voice.

It didn't take long to get a sense of what he was talking about.

Just a few days after my meeting with Bart, more than 500 B Corps publicly committed to reaching net zero emissions by 2030. Embedded in that commitment was a requirement to focus strongly on emissions reductions at source (as opposed to through offsets), as well as getting started on cutting those emissions now (as opposed to waiting until 2029 to get rolling). Crucially, the B Corp Climate Collective—as the group had named itself—timed the announcement as part of the COP25 climate talks in Madrid. In doing so, the goal was not simply to demonstrate these businesses' superior practices or corporate values, but rather to put pressure on other leaders to step up.

Even though there was plenty of focus on getting each signatory's own house in order, the Collective's statement that accompanied the pledge was unequivocal that these efforts were a precursor to, not a replacement for, wider political change:

> We believe, therefore, that it is imperative for all businesses to demonstrate leadership in eliminating emissions, drawing down carbon, and ensuring a just transition for displaced workers and communities to a net zero emissions economy. In addition, we believe it is imperative to use the

> power of our collective voice to advocate for policy changes necessary to remove impediments and align incentives that will drive meaningful climate action.

Soon after, we at Redwoods joined the Climate Collective too. As Yvon Chouinard once wrote, "Actions build on one another."

Beware the Benign Benefactor

Of course, the idea of corporate activism or advocacy really is nothing new. And it's also very far from being automatically benign. As we saw in Chapter 6, Big Oil and other fossil fuel interests have spent unimaginable amounts of money to hold back progress on climate and the environment. So while it might be tempting to uncritically welcome corporate voices speaking out for climate action, it's important to temper any enthusiasm with a cautious and detailed reading of what a business is actually advocating for.

It's hard to argue, for example, with a clothing company like Patagonia advocating for people to buy fewer clothes. But when oil companies start pushing for a price on carbon—as many have recently done—it's time for activists and legislators alike to start reading the fine print. Here's how Umair Irfan, writing for *Vox*, burst the bubble of naïve optimists like myself when Exxon came out in favor of a carbon tax:

> Cities across the United States are currently suing oil companies to make them pay for damages wrought by climate change, which could put companies like Exxon on the hook for billions of dollars in payouts.
>
> The proposal Exxon wants to enact is one that would shield the company from lawsuits while also preventing further climate change regulations. All in all, it would grant oil companies the kind of immunity from litigation the gun industry currently enjoys.[4]

Proposed immunity was by no means the only problem with Exxon's preferred solution. The other was, quite simply, that it was deliberately designed to not actually work.

At a proposed price per ton set at $40, the plan was way too modest to have any real impact on emissions. In other words, while other countries were exploring an outright ban on cars with internal combustion engines, here in the States, Exxon wanted to make fueling such an engine just a little more expensive, and to do so over a painfully long period of time. In the same *Vox* article referenced above, Joseph Majkut, director of climate policy at the Niskanen Center, summed up the basic rationale in very simple terms: "It prevents other things that they think would be worse for their business model."

If businesses really are serious about advocating for climate action, then they are going to have to be prepared to advocate for real, robust solutions at the legislative level—even and especially when those solutions are not in their immediate, short-term financial interests. This challenge gets trickier the closer a company is to the current, fossil-fueled hegemony.

Capitalists Against Unbridled Capitalism?

As we saw in Chapter 6, the likelihood that Exxon or Shell will stop opposing, let alone become advocates for, a genuinely zero-carbon system is almost vanishingly small. Yet there are plenty of questions about other sectors of the economy too:

- Will legacy car companies go all-in for electrification, not to mention moving beyond the whole concept of private car ownership?
- Will the aviation industry support regulatory measures that would shift long-distance transport to more genuinely sustainable options?
- Will food conglomerates move beyond niche organic and plant-based product marketing, and instead embrace policy reforms that prioritize regenerative agriculture, carbon farming, and other forms of planetary repair?

There's good reason to believe that, for many businesses, the answer will be a resounding "LOL no." Yet the moment we are in demands nothing less.

Kim Coupounas—global ambassador for B Lab, the non-profit that certifies B Corps—has been instrumental in spearheading that movement's climate advocacy efforts. In a video call to discuss the work of The B Corp Climate Collective, she did not hold back in communicating exactly what was at stake:

> The only way to solve the level, and scale, and depth of the challenges in front of us, is to step into the ring with everyone and do your part. However far you can go and still maintain your survivability as a company, and as an individual, that's the kind of scale of change that's needed now. I don't want to sound too radical, but we have nine years to halve emissions before all hell breaks loose.... We haven't seen anything yet.

If Kim is right about the urgency of the situation—and the science would strongly suggest she is—then businesses are absolutely going to have to get real about cutting their own emissions. But they can't stop there. They are also going to have to learn to build on what they are doing in-house, and use it to demand change that cascades across the entire economic system—up to and including a willingness to question some of the central tenets of profit-at-all-costs capitalism itself, such as perpetual economic growth. Ironically, they may also have to use their voice to challenge the power of corporations in our democracy. And we, as citizens and as consumers and as employees, are all going to have to get better at holding them to account for doing so.

After all, we are not short of problems that urgently need fixing.

8

Swimming Upstream

The start of the journey was actually pretty glorious.

My close-to-seventy-year-old mum had recently gifted me a Blix e-bike. (Mostly, this was an excuse for her to have one available when she visited us in the United States.) Deep green, heavy as hell, with a step-through frame, it had a sit-up-and-beg riding position and a sturdy basket on the front. While my buddy Matt made fun of me every time he saw me on what is, quite literally, an *omafiets* (a charming Dutch term that translates as "granny bike"), I personally loved its classic European looks. It even had a cup holder on the front should you need to pick up a latte on your way to work.

Offering power assist from a small lithium battery, the bike had already obliterated many of my excuses for not biking around town. I had started out with the occasional grocery trip, or a car-free jaunt to meet friends for a pint. This rediscovery of cycling now had me wondering what it would be like to ride the 14 or so miles to the offices of Redwoods, which I was working for at the time as an independent contractor. So that's how I found myself walking my daughters to school one morning, hopping on my trusty green steed, and whizzing through the rush hour of Durham.

"You Are Definitely Going to Die"
What many people unfamiliar with e-bikes don't realize is that much of the benefit does not come simply from avoiding the physical work of cycling. (Although, I confess, that is an added bonus.)

Rather, the addition of a small, motorized boost allows you to better keep up with traffic, and empowers you to hold your own in car-centric traffic situations. So as I pulled ahead of the cars at a set of traffic lights, and navigated my way toward seven miles or so of car-free greenway known as the American Tobacco Trail (ATT), I was feeling confident in my chosen mode of transport.

After about seven miles on the ATT, however, the real challenges began. First, I wobbled my way down the two-lane Cornwallis Road, trying as best I could to stick to the narrow, painted bike lane that hugged the curb. Then I turned onto the much wider Davis Drive, and traversed my way into the middle of three lanes. Eventually, I managed to get all the way over to make a left-hand turn as SUVs, minivans, and cars came streaming from the off ramp of I-40.

I began to wonder if this had been a good idea.

The Research Triangle Park (RTP) sits next to Raleigh-Durham Airport. It covers 7,000 acres, comprises 22,500,000 square feet of built space, and is home to corporate giants that include Cisco, SAS, IBM, as well as the National Institute of Environmental Health Sciences. While bigger than many, it's the kind of office park landscape that would not be out of place on the outskirts of any major North American city. I had been working sporadically from the Redwoods offices there for some time, typically making the commute in my Nissan Leaf a couple of times a week. If this morning's e-bike expedition succeeded, I was considering making it a more regular habit. Yet as the large, hulking automobiles rushed by me, the doubts my wife had expressed to me that morning ("You are definitely going to die!") were starting to feel just a little bit too real.

Eventually, however, I did reach my destination. I locked my bike to the always-empty bike rack outside, grabbed my morning coffee, and plugged the removable battery in to recharge, already nervous about the afternoon journey home. On receiving a few inquisitive looks regarding my helmet, I explained what I had been up to and asked if anyone else ever rode to the office: "Sure, I think

Rich over in underwriting used to ride occasionally. He stopped when he was knocked from his bike and broke several ribs."

Meeting People Where They Are

For all the ink that has been spilled offering tips, suggestions, and encouragement for lower-carbon lifestyles, these articles have tended to spend too much time focused on one-size-fits-all pronouncements, and too little time acknowledging the fact that certain lifestyle changes are easier for some than they are for others.

That's the thing about personal choices. We don't get to make them in a vacuum.

For some, giving up flying might open up fantastic new adventures and save thousands of dollars. For others, it means compromising your career, spending *more* money, or never seeing family again. We all face similar tradeoffs in almost any aspect of lower-carbon living. Whether it's going vegan, choosing a different job, donating your time or money, or changing your consumer habits, we navigate a constant balancing act that includes our own personal values and desires; the values and desires of our family and loved ones; the expectations of bosses, co-workers, or the culture in which we live; and the influence of structures and systems that reward one choice over another.

The nature of those tradeoffs will depend greatly on where you find yourself in the world. It's easy enough for folks to take the train between Edinburgh and London, for example, or between Tokyo and Shin-Osaka on the 178 mph bullet train. But to travel by train from Durham to see family in Indianapolis would take me a solid 35 hours and cost hundreds of dollars more than an airline ticket or driving. It would also involve burning a very large amount of highly polluting diesel. Consequently, even if I do personally decide to make that "sacrifice" (and I have greatly enjoyed my few train trips from here to New York City), I do so while fairly confident in the knowledge that not many of my neighbors are likely to follow. And thus it starts to feel like an exercise in futility.

On the flipside, it's super easy for me to walk or e-bike to my nearest grocery store or restaurant and pick up some organic food or plant-based meat alternatives. That's not the case for my more rural neighbors. Nor is it viable for folks living on the other side of Durham—where there are entire neighborhoods where the only realistic grocery option is the local convenience store or gas station, and where biking, walking, and mass transit use are considerably less convenient or safe than in my more affluent, predominantly white neighborhood.

All this might seem like stating the obvious. After all, it goes without saying that eating healthy food is easier if you have access to stores and money to spend. Likewise, walking is easier if you live near your destination. And, of course, biking is a dream if your streets are designed with the cyclist in mind. So far, so repetitive. Yet for far too long, the focus on voluntary behavior change and lifestyle "choices" has ignored the fact that those choices are often not really a choice at all.

We have to meet people where they are. And where they are is often—environmentally speaking at least—somewhat of shitshow.

Changing the Direction of the Current

In December of 2007, filmmaker Annie Leonard released an animated online video called *The Story of Stuff*. In that video, she revealed the hidden connections between the products we buy and their negative impact on society. That initial release quickly notched up 12 million views and sparked conversations about the long-term consequences of globalization and consumerism. Welcome as those conversations were, it was no surprise when the gravitational pull of our consumer culture quickly drew the discussion toward changing our shopping habits, rather than changing the system that makes those habits the default option.

In 2012, however, Leonard released a follow up video called *The Story of Change*. This particular video set out to elicit a more sophisticated response. Describing ethical shopping as "a great place to start but a terrible place to stop," she explored lessons from social movements throughout history, and concluded that real change can only come about through developing a bold vision and then putting pressure on authorities to make that vision a reality. The trick, she said, was to shift our primary focus from flexing our consumer muscle to flexing our citizen muscle:

> Trying to live eco-perfectly in today's system is like trying to swim upstream when the current is pushing us all the other way. But by changing what our economy prioritizes, we can change that current so the right thing to do becomes the easy thing to do.

Trying to swim upstream seems like an apt metaphor for my attempt at bike commuting. It's not that the task itself was impossible. The conditions had been set, however, to make it inconvenient, unpleasant, and unlikely to become the norm for anyone but the most deeply committed. And while it's perfectly possible that my e-bike adventure helped a few people visualize alternatives, part of me wondered whether my precarious wobbling might be having the opposite impact to what I had hoped. I can easily imagine, for example, an SUV-enclosed family driving by, and gawping incredulously at the suicidal fool who was taking such questionable risks.

At the time of writing, I have not yet attempted that journey again. I do, however, have good intentions of trying.

Modeling What's Possible

Peter Kalmus is more familiar than most with the art of swimming upstream. In fact, he has come to quite enjoy it. A NASA climate scientist living in suburban Southern California, he has gone to what many would consider extraordinary lengths to cut his own impact on the environment— bringing his carbon footprint down to something close to one tenth of the US average. He documented these efforts in his 2017 book called *Being the Change: Live Well and Spark a Climate Revolution*.

When I called him up to discuss this work, he shared that he hadn't always been focused on the climate crisis. In fact, he pointed to two specific turning points in his life which led him to an awakening of sorts. The first was the birth of his eldest son in 2006—an event which he says made him "less self-centered" than he might otherwise have been. The second was attending a lecture in the same year by famed NASA scientist James Hansen, in which he explained how radiative forcing was destabilizing the Earth's climate at an unprecedented rate. At that lecture, Peter recalls both learning about the emergency in terrifying detail, while simultaneously witnessing how few people—even within the scientific community—were really grasping the magnitude of what we are facing:

> The lecture had me on the edge of my seat.... This was a physics colloquium, so there were a hundred plus physicists, grad students, professors, and post docs in a lecture hall listening to Jim Hansen talk about the energy imbalance of the Earth, and I felt like for most of them it was just another physics lecture. Which was weird to me, because this is our planet and this is my kids' futures.

This disconnect between what the scientific community was discovering and communicating, and what it was actually doing to fix the problem, led him to what he describes as the "long, messy journey" of figuring out how to be as effective as he possibly could as an activist. He eventually came to the conclusion that if the crisis was as urgent as the science suggested, then scientists might need to lead the way in demonstrating how badly things need to change:

> It's good for climate advocates—especially those who have pretty decent sized platforms—to take reasonable steps to reduce [their emissions]. This really is an emergency. That's a personal judgment I have, based on the science and the projections now. But I believe it is a completely reasonable judgment. And if something's an emergency, you don't carry on like nothing is wrong. The whole definition of an emergency is that it makes you stop whatever it is you are doing and you put your response to that thing as the highest possible priority.

What that looked like for Peter was a pretty concerted effort to tackle his own carbon footprint. He converted an old diesel car to run on waste vegetable oil. He pledged never to fly again. He transitioned his family to a vegetarian diet with a heavy dose of dumpster diving (rescuing expired food from the grocery store waste stream). He even embraced 'humanure', building a system for composting his toilet waste in his suburban California home.

While he acknowledges that his lifestyle may not be for everyone, he is careful to emphasize that he has found much joy in living in a more intentional, low-impact way. And he is confounded by the idea that anyone would want to discourage someone else from pursuing such a path:

> I would urge all activists to at least give it a try, for several reasons. One, they might like it more than they think. And the second, main reason is that it might make their voices more powerful. I would never have published a book at this point—or really had any kind of platform at all—if I hadn't made those changes to my own path. That was a key part of finding my voice. And if you don't find it useful, then don't carry on—but also don't try to discourage those of us who are on this path. There are so many different ways to go about climate activism, and so many things that might bring people to the movement.

Peter is under no illusion, however, that individual behavior change is going to deliver the kinds of emissions cuts that the climate crisis demands. Instead, he argues that it's through modeling alternative ways of being that we are best positioned to encourage societal-level action. It is also, as he commented to me over the phone when I shared my e-bike commuting story, a way that we can learn more about exactly how the system shapes our choices.

Just as my former editor Lloyd Alter had demonstrated with his own efforts to live a 1.5 degree lifestyle, Peter too suggested that it's where we feel the most friction in our personal efforts that we need to focus our demands for systems-level remedies:

> By taking this journey you really start to see how you are
> limited by the systems, and how important it is for the sys-
> tems to change. There's this really deep connection between
> your own reductions and your awareness of systems change.

Peter is quick to point out that he is not pushing for perfection. In
fact, frustrated by one too many Twitter discussions about systems
versus individual change—a framing he rejects as a patently false
choice—he was actually a little hesitant to talk to me at first when I
told him I wanted to talk about my "defense of eco-hypocrisy." (The
original working title to this book.)

As we talked, however, it was clear that we were largely coming
to the same conclusions: That individual action matters more as
a path to influence and cultural impact than for its ability to radically cut carbon in and of itself. Part of the trick, he suggested,
was to focus on the joyful, enriching aspects of the journey, so that
others can see the benefit in making a change:

> The reason I wanted to write a book was to get the message
> out that this isn't all about sacrifice—there's a lot of awe-
> someness about burning less fossil fuels.

Even for folks like Peter, however—who has gone further than
most in reducing his impact—there's a diminishing return on investment as you get closer to zero impact. Not only does it get
harder to cut emissions as the low-hanging fruit is plucked but,
Peter argued, there's also a concern that getting too far out in front
means that fewer people will follow. Peter described his approach
to solving this conundrum as being a little bit like a balancing act
or dance:

> I was still driving a bit and using gas to heat my home, and
> burning about two metric tons of CO_2 as a result. But the
> stuff I would have to do to go further would be really hard.
> So instead I put my energy into activism, and writing this
> book. And what I realized was that was actually a good

> thing—because more people could follow me and it was more relatable.... It's a little like surfing on a wave. You want to be pushing a little further than where the system is at.

Peter's ultimate goal is not, and never has been, to wipe his personal carbon slate clean. Instead, it's to start changing what's possible in terms of politics. And here, as Peter says in his book, we have to follow the money:

> Burning fossil fuels imposes huge costs on society that aren't included in the price of fuels, primarily by causing global warming and respiratory illness. It's crucial to fix this market failure because few of us will voluntarily stop burning fossil fuels in a society that still strongly rewards this behavior.[1]

In other words, we have to change the rules of the game.

Subsidizing the Incumbents

Talk to your average commuter about the car-centric nature of many American communities, and you'll likely hear that it is a result of basic market forces. Americans like to drive, so America builds roads. That, however, is an oversimplification. Sprawling developments like the Research Triangle Park—where my somewhat precarious bike ride took place—have come about through a confluence of forces that time and again have served to put oil and automotive interests above the health and wellbeing of ordinary citizens.

Chief among those forces have been the massive, direct subsidies afforded to oil, coal, and gas companies. According to environmental pressure group Oil Change International, between 2016 and 2017, the US spent an average of at least $20.5 billion a year on direct subsidies to fossil fuel industries. In other words, taxpayer money was going to prop up a powerful incumbent, thus keeping the cost of fossil fuels low and artificially penalizing the alternatives.

Huge as they are, direct subsidies are just half of the story. Whether it's the premature deaths related to asthma and heart disease, the more than 35,000 people killed on our roads every year, or the $81 billion in military spending focused on defending oil supply routes,[2] there are very real costs involved in the consumption of fossil fuels that are never seen in the prices at the gas pump. And that's before we even get started on the economic impacts of climate change. But how much, exactly, would you be paying at the pump under a more realistic pricing model?

Stefan Tscharaktschiew, an economist at the Dresden Institute of Technology, has calculated that if we added up all of the social and environmental costs of gasoline consumption in Germany, drivers would be paying an optimal tax of $4.36 a gallon over and above the market cost of the gas itself.[3] Meanwhile, a 2019 paper from the International Monetary Fund calculated that once the social and environmental impacts of fossil fuels are fully accounted for, the real figure for both direct and indirect subsidies amounted to $5.2 trillion in 2017, or a whopping 6.4 percent of global GDP. Had these costs—which are most likely out of date and underestimations by now—been fully factored into the price of coal and oil in 2015, the paper's authors suggest that this pricing correction alone would have led to a drop in global carbon emissions of 28 percent, as well as 46 percent fewer deaths from air pollution.[4]

It's no accident, then, that I was surrounded by gas-guzzling SUVs on that day as I rode to work.

The Destructive as the Default

Similar observations can be made for almost every aspect of our status quo. If consumers were asked to pay not only for the direct costs of raising cattle, but also a dollar figure to reflect the societal costs of methane emissions, deforestation, soil erosion, water pollution, not to mention the negative health impacts of excessive red meat consumption, then it's fair to say that our dietary habits would change dramatically.

Even a partial accounting for so-called externalities would radically transform the economics of food production. As just one example of how this might quickly add up, a 2017 study by non-profit group GRAIN found that the world's top 20 meat and dairy producers alone emitted 932 megatonnes of greenhouse gas emissions, which puts them right up there with entire nation states in terms of the damage being done:

> If these companies were a country, they would be the world's 7th largest greenhouse gas emitter. It's now clear that the world cannot avoid climate catastrophe without addressing the staggering emissions from the largest meat and dairy conglomerates.[5]

Any serious effort to address climate change can and must include an effort to have the price of a product reflect the true cost. But tackling pricing signals requires generating political will. And political will can be hard to come by if citizens will be asked to pay more for their burgers and their gasoline.

Clearly, we are going to have to change our culture too.

Writing a Different Story

Rachel Malena-Chan describes herself as a settler on Songhees and Esquimalt Peoples' lands. She is a Canadian citizen, climate activist, and communications strategist, now based in Victoria, British Columbia. We first started talking when she launched EcoAnxious .ca—an online platform where those concerned about climate change can share their fears and explore their anxieties without judgment or preaching. I reached out to her by phone to try to understand the thinking behind this approach.

Rachel grew up in a religious environment that encouraged philanthropy and altruism, but it wasn't until her twenties that she translated that upbringing into the idea of fighting for justice on a more structural, societal level. It was during this awakening that she was exposed to the ideas of the academic and activist

Marshall Ganz, whose work on the role of storytelling in social change has influenced everything from the civil rights movement to the Obama election campaign.

When she started a Masters in Community Health at the University of Saskatchewan, she picked up a copy of Naomi Klein's *This Changes Everything*. She quickly realized that climate and environment was a big missing piece within the social justice movement:

> I kept coming across this question: is it really this bad if no one is really talking about it? Or, on the flip-side, is it really that people don't know enough or care enough, and that's why nothing is happening? Neither of those things really made sense to me. I was pretty sure I was correct in terms of how serious it was, or how far gone the situation was already.

To put it another way, if people knew and cared about the issue, why were they not mobilizing to do something about it? Rachel began to apply the work of Marshall Ganz as a framework for understanding this question. As part of that work, she conducted interviews with social justice activists. She found that many interviewees knew all too well that the climate crisis is a real and immediate emergency, yet this knowledge didn't necessarily translate into changes in their personal lives. The way she described it to me sounded more like a play, or a movie, than a traditional model for raising environmental awareness:

> The way I see it, the real challenges come in Act One, at the beginning of the story, when people realize how difficult this is and how bad things are. That's true for people who have been in this work a long time, and also for people who are new to the story.
>
> Then the other big point of challenge comes in Act Two, in the middle of the story, when we realize we are going to need to be courageous without knowing what the future holds.

> And finally we get to Act Three, where we have a need to know that what we are doing is meaningful—that it will have a genuine impact on the issue we are trying to address.

As I talked more with Rachel, I came to understand that her talk of stories and acts and narrative to describe life choices and life experiences was not simply metaphorical. "Stories," she explained, "are as fundamental to the human experience as mass is to the physical world."

In other words, it is only through rewriting our stories that we can hope to engage our culture on any meaningful level. And that means leaving room for nuance, depth, and complexity. After all, a story where every character is "the consumer" is not going to inspire us to create lasting change. Similarly, a story where we halt climate change by voluntarily slashing each of our carbon footprints to zero is equally hard to believe in. Rachel explained to me how viewing her own lifestyle through the lens of storytelling has allowed her to look at personal choices from a slightly different, more forgiving, and yet deeply values-led angle:

> My travel choices have definitely shifted. I do still fly, but I think quite critically about why and where I am flying. And I try to make sure that resonates with the things that are meaningful to me—including spreading the word about the work that I am doing, equipping others to do this work, and of course seeing family and friends.

A More Interesting Conversation

Interestingly, Rachel says that this consciousness around emissions and impact has led to conversations with family and friends, some of whom have talked about moving closer to each other to reduce the need for travel. And it's within this space of discussion and storytelling, rather than blanket pronouncements or judgment, that Rachel has found constructive opportunities to explore alternatives:

> Truly, if I did or didn't buy a single flight would not meaning-
> fully change the trajectory of the aviation industry. So I've
> had to come to terms with this from both sides. Choosing to
> buy a flight is inconsistent with my values. And yet saying
> that I'm the reason the world is going to end because I saw
> my family for Christmas is also a lie. Figuring out the long-
> term plan is a more interesting way to come at it.

The focus, then, is not on each individual decision in every specific
moment. But rather, what needs to happen as we navigate a path
down from our current highs, and how do we, as individuals, best
fit into that story as effective agents of change:

> "When are you going to start planning for this?" is, I think,
> a decent way to come at this for those who are just starting
> out. And for those of us who aren't just starting out, to say,
> "Are we really being honest with ourselves about the lon-
> gevity of this behavior, and what is our plan going to be
> when things do finally begin to change?"

By thinking of ourselves as characters in a narrative, and by
grounding that narrative in the moment we find ourselves in, it becomes possible to think more critically about where we invest our
energy. That means accepting that many of us who are trying to
make more responsible choices are part of what Rachel describes
as a "Transition Generation," which was born on the cusp of major
change, but still has a foot on either side of the divide.

That does not mean washing our hands of responsibility or giving up on change. It means focusing our efforts on where change
is most likely to happen.

9

Focus, Goddammit

On April 16, 2014, a group of South Korean high school students were on a field trip from Incheon to Jeju Island. They were traveling on the Sewol ferry, along with approximately 200 other passengers and 30 crew. Right around 8:45 in the morning, crew members on the heavily overloaded ferry made a series of steering errors—causing the ship to list violently and eventually capsize. While distress signals were sent out almost immediately, it took several hours for the coast guard to arrive. During that time, victims were told to shelter in place inside a clearly sinking vessel. It later emerged that the coast guard never even entered the ship.

In total, 304 victims lost their lives that day. Tragically, the death toll included 254 seventeen-year-olds from the high school group.

The South Korean public was outraged at the unnecessary loss of life from a botched government rescue, and initial news reports and social media posts placed much of the blame on the shoulders of President Park Geun-hye. Unsurprisingly, the president's approval ratings plummeted as a result, falling from a high of 70 percent well into the 40 percent range in the weeks that followed.[1]

Geun-hye's regime was not willing to take the accusations of inadequacy sitting down, however. As revealed in a Secret Service report that was released in November of 2016,[2] the government undertook a deliberate and concerted campaign to manipulate public opinion—primarily by shifting attention to the actions of the crew, and of the ship's owner Yoo Byung-eun. (Yoo Byung-eun would later be found dead during a nationwide manhunt.)

An Effective Exercise in Distraction

Analyzing both media and social media discussion in the four months following the Sewol disaster, S. Mo Jang of the University of South Korea and Yong Jin Park of Howard University, have shown a clear and inverse correlation between competing storylines:

> The more the media and public focused on the ship's own-ers, the less attention was paid to the bungled response of the Park regime. And the more the attention was on the Park regime, the less oxygen there was for any discussion of the part played by the ship's owners.[3]

As with any such tragedy, an objective analysis of the situation would likely suggest that there was plenty of blame to go around: regulatory missteps that led to an unsafe operating environment, a poorly orchestrated rescue by the federal government, a grossly overloaded ship, an owner who prioritized profits over safety, a captain who was away from the helm, and a crew that managed to save themselves while leaving hundreds on board to drown. And yet despite the multitude of culpable parties and contributing fac-tors to the disaster, the public discourse proved unable to hold its attention on multiple storylines at once.

This isn't exactly a startling revelation. And yet it's important.

Attention Is a Limited Resource

Politicians and the powerful have always used distraction and scapegoating as a ploy to evade culpability. The study of the Sewol ferry disaster serves as a poignant reminder that both media cover-age and public attention are not inexhaustible resources. It is criti-cal that we keep this fact in mind when we talk about the climate crisis, or any other complex, systemic issue for that matter.

As we've already seen throughout this book, there is a strong case to be made that the *systems change versus lifestyle change debate* is in many ways a false choice, or at least an oversimplified one. There is nothing wrong with riding your bike, eating vegetarian, or giving up flying. Adopting any one of these lifestyle changes contributes to reducing emissions, and reducing emissions is exactly what has to happen if we are going to make our way through this crisis.

Yet there are opportunity costs involved whenever we spend time talking about one thing at the expense of another. Focusing our advocacy on lifestyle change—and specifically on encouraging individuals to reduce their personal carbon footprints and their individual environmental impact—does run the risk of distracting attention away from the scale at which workable solutions are likely to emerge.

First Things First

Few people within the movement understand the need to focus both our resources and our attention better than Mary Anne Hitt. Formerly the executive director of Appalachian Voices, she fought against mountaintop removal coal mining in her native West Virginia for years. While she was initially motivated by a desire to save the local landscapes she loved, engaging in that fight led her to recognize the destructive nature of the coal industry in a much more systemic way.

As she told me during one of the more energetic discussions I have had via webcam, the sheer scale and diversity of problems caused by coal made it a natural target for environmentalists to focus on. Whether it was air and water quality, habitat loss, climate, or waste (the coal industry had the second largest waste stream in the country at the time), there really was something for everyone in terms of motivation for demanding change. And because coal power was—in the early 2000s—the single largest source of greenhouse gas emissions in the US, it made sense for activists to go head-on at taking it down first.

During the course of her work with Appalachian Voices, she began to fight a proposed new coal plant in southwest Virginia— and through that process learned that the George W. Bush regime was planning the construction of more than 200 new coal-fired power plants nationwide. Her organization joined a growing coalition of grassroots and national environmental groups named Beyond Coal. Coordinated by the Sierra Club, Beyond Coal was laser focused on bringing communities together in order to defeat this proposed expansion of coal use in the US, which they saw as a potential "game over" scenario for the climate fight:

> The Beyond Coal Campaign was born out of a desire to focus. It was very clear that if those 200 plants were built they would lock the US into 50 or more years of our dirtiest possible source. Once they were built, utilities would want their money back—meaning there was little chance of reversing that damage.

As I told Mary Anne during our chat, I have a distinct memory of feeling profoundly and prematurely defeated at the mere thought of what these proposed coal plants would mean for the climate. And I also have a distinct memory of hearing about the launch of Beyond Coal and feeling skeptical that they could ever achieve their goals. After all, at the time, industry watchers were projecting coal use to expand not just in the United States, but exponentially worldwide, with China and India leading the charge.

"I honestly didn't think y'all could pull it off," I told her frankly. Mary Anne laughed. "We weren't sure we could pull it off either."

Yet they had no choice but to try.

The Beginning of the End of Coal

Ultimately, the campaign proved wildly successful. Of the 200 proposed plants, the campaign managed to halt 170 of them. (When we chatted in May of 2020, Mary Anne was actually fresh from celebrating the defeat of the last of these plants just a few weeks before.) And once it became clear that the lion's share of *new* plants was no longer going to be built—around 2009 or 2010—the campaign shifted its focus to shutting down the 530 *existing* coal plants that were already in operation across the nation. (Mary Anne left Appalachian Voices and became the director of the Beyond Coal campaign right around this time—before more recently moving on to become director of campaigns at Sierra Club.)

At the time of writing, Beyond Coal has now contributed to the retirement of 307 of those 530 existing plants. And while they still have 223 left to go, Mary Anne boldly declared that she fully expects the US to have effectively eradicated coal from the electricity grid by 2030, if not before.

Given that the Beyond Coal campaign greatly exceeded the expectations of many observers like me, I asked Mary Anne whether she thought its singular focus made it so successful. I also asked her whether this offered a potential path forward for the broader climate movement. After all, even though there are many different sources of emissions and causes of global climate change, research has also shown the vast majority of emissions can be pinned to a few sectors of the economy, and—from the production end at least—a very small number of large corporations and state-owned entities.

Rather than asking would-be climate activists to do everything at once ("go vegan, ride a bike, take shorter showers *and* take on the might of global fossil fuels"), maybe we need to concentrate our messaging and prioritize our asks. She agreed that there was a lot of truth to this observation:

> We do only have so many hours in the day, and people need an entry point into the discussion. Beyond Coal provided that entry point for many people, including folks who may have been more concerned with local water or air quality issues than they were climate or energy policy.

She also paused, however. While she agreed that the climate movement does need to prioritize its actions, maintain its discipline, and focus its messaging, she also argued that it is critically important that we do not oversimplify the issue. The sheer complexity of the crisis, combined with the many intractable social issues that play into it, means that no single solution is ever going to be enough. We need to be able to walk and chew organic gum at the same time:

> You do have to be careful though. There is a danger to putting your blinders on and not seeing the bigger picture. I get frustrated, for example, when people want to tackle climate change without also addressing social justice.

As we continued our conversation, we zeroed in on the competing pressures inherent in working on multiple things at once, while also accepting that none of us can do everything. Mary Anne noted that while this is true of campaigning, it is also true of what people choose to do in their own lives.

It's good to live your values and seek points of leverage within your lifestyle, she said, yet the problem is much too vast for any one individual. That means it's not necessarily helpful to pursue eco-perfection, especially if it drains your energy from focusing on a specific and strategic piece of the broader puzzle. It can even be downright unhelpful if it excludes others from joining your efforts.

Fortunately, none of us are ever really acting alone.

Being "Better"

My friend Mary Annaïse Heglar—who I interviewed earlier in the book—once showed up to a virtual board meeting of Dogwood Alliance wearing a t-shirt emblazoned with the word "#vegan" in large white letters. When I complimented her on her shirt, she looked a little embarrassed. She had actually forgotten she was wearing it, and confessed that she would never leave the house with it on: "I only wear it with people like you. It would be way too obnoxious to walk around town in."

Vegans have gotten a bad rap for being sanctimonious, and it's certainly true that some will not shy away from a food fight. Yet, in my experience, this is the exception rather than the norm. Many of the vegans in my life look on their own dietary choices as just that—their own—and would much sooner enthuse about the delicious wrap they had for lunch than they would judge someone else for what is, or what is not, on their plate.

Deserved or not, however, the reputation for piousness has led folks like Mary to feel decidedly uncomfortable, or even shy, about displaying their dietary tendencies on their sleeve. So why is that?

According to researchers Cara C. MacInnis and Gordon Hodson, vegans actually experience bias or stigma to the same level as individuals suffering from drug addiction.[4] In an article for the BBC,[5] journalist Zaria Gorvett suggests that peoples' animosity towards vegans often stems from a rather surprising source.

While some people disagree with vegans on an ideological level, for others, their resentment actually originates from the fact that their own personal values are largely aligned with veganism. It's well documented, after all, that a growing number of us are coming to understand the catastrophic impact of excessive meat eating. Yet it's also clear that we aren't quite ready to voluntarily change our ways or give up eating animal products entirely. According to Gorvett, it's that disconnect that contributes to the tension:

> There's mounting evidence that we're particularly threatened by people who have similar morals to us, if they're prepared to go further than we are in order to stick to them. In the end, our fear of being judged far outstrips any respect we might have for their superior integrity.

This insight has interesting implications for the debate around behavior change. After all, if people are as likely to recoil or react angrily to examples of people who are being "better than them"—especially on issues they 100 percent agree with—then simply living a greener lifestyle and urging others to do the same may not be enough to convert new recruits to our cause. (Some readers may remember the South Park episode in which a hybrid car emits deadly clouds of smug.)

But what if the power of a vegan doesn't lie in converting other people to veganism? What if, instead, their power lies in gradually transforming menus? In that effort, vegans actually have a growing number of allies.

Meat Eaters and Vegetarians Unite

In May of 2010, my former boss Graham Hill—founder of *Tree-hugger*—appeared in an online video hosted by TED (of "Ideas Worth Spreading" fame) in which he talked about his identity as a "Weekday Vegetarian." It was, he explained, the culmination of a long journey that started with asking himself a fairly basic question: "Knowing what I know, why am I not a vegetarian?"

This question stemmed from Graham's growing awareness of the environmental destruction caused by industrialized animal agriculture. And it also stemmed from his own unwillingness to go all the way and give up meat entirely. (He had been a vegetarian in his past.)

Rather than continue to stress over an all-or-nothing approach to eating animals, he decided instead to pursue a different path— eat "nothing with a face Monday through Friday," and then give himself pretty much free rein on the weekends. There was some cheating here and there, and there were also plenty of veg-heavy meals on the weekends, but this middle ground approach provided a structured, easy-to-follow way to still make a difference.

The idea struck a chord. The TED video in which he introduced the concept to the world has now been viewed more than 2.7 million times, and Graham's weekday vegetarian concept spawned a series of popular recipes and posts over at *Treehugger*, as well as a book. He told me over a Zoom call from his house in Hawaii that his intent was always to create a more inclusive space where everyone can feel like they are making a difference—thus building a movement that has the potential to actually scale:

> If all of us ate half as much meat, then it has the same impact as half of us being vegetarian. It's hard for me to envision everyone going veggie, but half as much meat…we've basically already been there. We were there in the '50s. The amount of meat on our plates has grown from being kind of a side thing to pretty much taking over the meal.

The popularity of Graham's TED talk was just one sign among many of veg-centric eating going mainstream throughout the 2010s. Yet even as countless celebrities—including famous bat-biter Ozzy Osbourne, as well as athletes like Venus Williams—came out as vegan, studies suggested that the number of actual, self-identified vegans and vegetarians had remained remarkably steady.[6] While most of us weren't ready to commit to 100 percent, however, there was a growing contingent willing to take some significant steps in the right direction. In fact, according to one study in 2019, as many as 33 percent of US households reported at least one member of the family making efforts to eat less meat.[7]

Alongside Graham's Weekday Vegetarian concept, various terms and strategies were being floated around. "Meatless Mondays" had been a thing for a while. Food writer Mark Bittman coined the concept of VB6 (Vegan Before 6 p.m.). And food systems expert Brian Kateman started promoting the idea of "Reducetarianism," founding a non-profit group called The Reducetarian Foundation, which now holds an annual convention focused on the topic.

While approaches and the degree of strictness varied, a central theme that ran through all of these concepts was similar: reducing the total consumption of meat within our food system was more important than promoting strict adherence to a specific set of rules for any one individual. As Brian Kateman told me over the phone— and as Graham Hill had argued too—ideas like reducetarianism or Weekday Vegetarianism offered a pathway to overcoming personal divisions ("vegans are reducetarians too"), and instead building a movement that focused its effort on the only scale that really matters:

> The Reducetarian movement was built around the premise of a world where a large number of people participate in reducing their animal products by a small degree, rather than focusing entirely on getting a small number of people to cut

out their consumption of animal products all together. The genesis of that was just math. If we can get a lot of people to do something small, that would make a quantifiably larger difference than getting a small number of people to do a lot.

Brian also pointed out that, in addition to simply allowing more people to participate, this shift of focus away from personal purity also appears to have contributed to a shift in strategies. Specifically, he said, he had noticed an increased focus on systemic and institutional interventions that were aimed at changing the environment in which all of us make our food choices, not just those who had explicitly subscribed to a philosophy of avoiding meat:

The Reducetarian Movement has, in fact, become a movement. It is no longer just comprised of individual people who are taking steps to cut back on animal products, but it also includes various institutions and groups who are thinking about how they can advance that societal goal.

These institutional responses started showing up in many different ways.

The System Responds

Burger King began selling the Impossible Burger—one of a new breed of hyper-realistic plant-based "meats" that were explicitly marketed to carnivores. (Some 70 percent of Impossible Foods' customers are, apparently, regular meat eaters.[8]) Companies like Greggs—a UK-based bakery chain—saw their stock prices skyrocket in response to the success of vegan sausage rolls. Ikea made an explicitly climate-centric commitment to increase the sale of plant-based foods. And, again citing climate change as a motivation, the co-working real estate company WeWork removed meat-based meals from all event menus, and even began refusing to reimburse employees for meals that included meat.

The lessons from this trend will depend on where you stand. On the one hand, you could argue that plant-based eating only became mainstream once its adherents stopped focusing on individual moral purity. On the other, you could say that this opportunity never would have opened up without vegans and vegetarians first forging the path. Either way, it was interesting to watch how a concept like what we choose to eat—something that is about as personal as you can get—eventually became more centered on scalable, systems-level reforms and interventions.

Which begs the question, where else might such cascading ripples of influence make themselves felt?

The Cheapest Way to Fry

Whenever the topic of carbon footprints comes up, it's a fair bet that someone is going to mention flying. And not without good reason. As Andrew Murphy, aviation manager at Brussels-based think tank Transport & Environment, explained in an interview with Umair Irfan for *Vox*, the decision to fly (or not) is one of the single biggest ways that individuals can increase (or not) their own personal contribution to emissions: "Euro for euro, hour for hour, flying is the quickest and cheapest way to warm the planet."[9]

Of course, the counter-argument remains that any individual's decision not to fly will have limited impact on emissions, unless it is accompanied by a similar decision by thousands of others. While "the plane was going to fly anyway" may sound like an abdication of responsibility to those who see flying through an explicitly moral lens, it is also a fairly logical reading of the situation. Whereas a decision to not drive on a journey will directly and immediately result in emissions reductions, the collectivist nature of air transport means that not flying will only make its mark if and when others are willing to follow.

Nevertheless, the simple fact that aviation allows humans to cover distances they would never be able to do otherwise means that flying is going to remain a high-carbon activity, at least for many decades to come. In fact, prior to the COVID-19 disruption to the industry, most experts were predicting that global aviation demand would continue to grow unless policies were put in place to stop it. (The International Civil Aviation Organization projected between 300 and 700 percent growth in emissions between 2005 and 2050.[10]) With the vast majority of the world's population never having boarded a plane (only 11 percent of humans took a flight in 2018), and with a mere one percent of the world's population—the frequent fliers—generating a whopping 50 percent of global aviation emissions,[11] there is also a strong equity and social justice case to be made for flattening and, ideally, reversing that growth.

Luckily, in some corners of the world, there are promising signs that that's exactly what's beginning to happen.

The Growth of *Flygskam*

Around 2018, the rise of school strikes and other forms of youth-led climate activism led to societal-level discussions in many countries around the impacts of aviation. "Flygskam," Swedish for "flight shame" (not flight *shaming*; it's something one feels within oneself, not a tactic for attacking others), became the subject of newspaper articles and television debates. And traveler preferences shifted as a result.

Swedish airports reported a four percent drop in passengers between 2018 and 2019.[12] German airports saw 12 percent fewer inter-city domestic travelers.[13] And all this was building on fledgling trends that had been evident in the previous decade: Train routes from central Scotland (Edinburgh and Glasgow) to London had seen a shift from 20 percent market share to 33 percent between 2005 and 2015.[14] Yet encouraging as these signs are, it's notable that they occurred in societies and across routes where viable alternatives already existed. And without viable alternatives, many individuals—especially immigrants or descendants of immigrants—find themselves in a pretty tricky dilemma.

An Inclusive Conversation?

Zakiya McKenzie is a journalist, academic, and nature writer. As a Black woman living in Britain, she has spent a great deal of time exploring how environmental issues intersect with those of race and class. She is now residing in my former home town of Bristol, England, and we initially connected via Twitter over a shared love of Marmite and good cheddar cheese.

The conversation quickly evolved, however, to address the fact that any debate over flying will need to take into account a diverse range of experiences and family circumstances. When we finally spoke in real time via webcam, Zakiya wasted very little time in arguing that aviation emissions are a particularly difficult place to start climate conversations with first- or second-generation immigrant communities:

> The conversation about flying just does not take that [the immigrant experience] into account. I don't think we can talk about Flight Free as this great moral and ethical thing to do without considering that a lot of people live in different parts of the world and they move around in order to keep alive. A wife or girlfriend or husband or mother might live in one country. A child might live in another. What are we saying to families when we say, "You should not fly to connect with the people you love?"

If I understood her position correctly, Zakiya was not arguing against the need to tackle aviation-related emissions as an urgent societal priority. Nor was she ignoring the fact that there are strong equity- and justice-related reasons to reduce aviation emissions globally. Rather, she was pointing out that focusing the conversation on the level of personal choices and individual sacrifice meant we were starting with the hardest part first.

Just like the topic of meat eating, however, there are promising signs that the conversation is beginning to move to the systemic level. Researchers at the University of California, Santa Barbara, for example, have estimated that up to a third of the university's carbon footprint can be attributed to academic conference travel and talks.[15] Academics and activists are now pointing to the sheer scale of that impact as an imperative for reducing institutional reliance on aviation. It's here, Zakiya argued, that the climate movement should be focusing its initial efforts:

> The Flight Free Campaign absolutely should be going after the universities that send people to the other side of the

> world for two days, where you spend more time in the air
> than you do at the location giving the talk that you went
> to give. So at the institutional level, and at the Big Bucks
> level—the folks who can afford to get in jets and not think
> twice—those conversations absolutely should be had.

Like many of the interviews in this book, we were holding our conversation during the COVID-19 lockdowns of 2020. Given the relative ease with which many conference organizers had gone virtual, Zakiya suggested that this was a perfect time to start thinking about the ways that our system rewards or incentivizes those who fly, while penalizing those who would prefer to keep their feet on the ground.

It's important to note, of course, that business-related flights account for only 12 percent of global passengers. That means that even a complete elimination of such travel would barely scratch the surface of aviation emissions. Yet those same passengers are typically twice as profitable as those flying for leisure,[16] meaning there's reason to believe that any institutional reductions in travel would have significant, non-linear ripples across the rest of the transportation system. Especially if combined with efforts to disincentivize other forms of frequent flying, such an effort would undoubtedly result in structural pressures that could allow alternatives (e.g., more sleeper trains, more low-carbon passenger ships, etc.) to emerge, thus making flight-free travel significantly more accessible for everyone.

Clearly, international travel in general, and air travel in particular, is going to be one of the trickiest challenges in societal-level decarbonization. Every single flight that is avoided gets us one step closer to where we need to be. Given the deeply interconnected, international world we live in, however, it seems reasonable to suggest that the best way to tackle this problem is by learning to see ourselves as parts of a more complex, and more interesting, whole.

It's a trick that's been done to great effect before.

10

What Difference Does It Make?

Visit Amsterdam Central Station today and you'll see three full stories of secure bike parking. Every day, hundreds of thousands of cyclists stream in and out of this busy terminal, following a set of unspoken and slightly anarchic rules that somehow keep everybody moving.

In so doing, they are continuing a long history of bike culture that dates back to the very early days of the bicycle itself. The scene is so integral to the image and fabric of Amsterdam that it's hard to imagine the city without the throngs of two-wheeled steeds now. However, the Amsterdam cyclist's position at the top of the city's food chain hasn't always been so securely assured.

Pete Jordan—an American transplant who moved to the city because of his love of cycling—has spent years documenting Amsterdammers' relationship with the bike. He compiled many of these stories into a book called *In The City of Bikes*. In that book he writes about the boom years of the 1930s, a decade that saw the addition of 10,000 bikes a year to the city's streets. And he digs deep into news reports and municipal archives to uncover the central role that bikes played in Dutch resistance to German occupation during the Second World War, including the large-scale bike confiscations that the Nazis and their collaborators inflicted in retaliation. ("Give me my father's bike back," is, apparently, a common insult still lobbed at German tourists today.)

These stories paint a picture of a city in which cycling isn't merely a dominant form of transportation, but rather an expression of civic identity. Ironically, it is a feature so ingrained in the culture that it often goes unnoticed by Amsterdammers themselves, even as outsiders and newcomers gush with surprise. A recent viral video showed a bike parking attendant of Kurdish origin expressing his admiration when Dutch Prime Minister Mark Rutte showed up to collect his ride: "This is how you run a country," enthused the parking attendant. "I am never leaving."[1]

There is nothing unusual about that scene. Whether it's the common sight of Dutch royalty, celebrities, and politicians riding in the streets, or the futile efforts of authorities to enforce order on the city's bike lanes, Amsterdam owes much of its unique nature to the fact that the bicycle has never quite been tamed. Yet it's not that people haven't tried.

In fact, if there is one thing that Jordan's book makes abundantly clear, there have been numerous and determined efforts to "modernize" the city and put auto traffic first. Jordan identifies the banning of bikes from Leidsestraat—a fashionable thoroughfare which was home to many of the city's most popular department stores and retail establishments—in 1960 as being a pivotal moment when the balance of power swung from the cyclist toward the internal combustion engine. And this was no accident. Even as cycling continued to be far more popular, the fact was that cars happened to hold more political and economic sway:

> ...virtually anyone who had influence over traffic policy (elected officials, high-ranking civil servants etc.) either owned a car or, as a job perk, had one at his or her disposal. While these well-placed, well-organized motorists "beat the drum to promote their own interests," the individual pedestrian or cyclist had little recourse for advancing their cause beyond penning a letter to the editor...[2]

Just as was happening in many cities around the world, this imbalance in political power resulted in an imbalance of power on the streets. Road layouts were changed, canals were filled in to make room for car parking, and citizens began to get the message. Between 1960 and the 1970s, the number of cars in city streets grew fourfold, and the number of cyclists willing to brave the onslaught dwindled as a result. Between 1965 and 1970 alone, the number of bikes entering the city each day dropped by 60,000.[3]

Organized Resistance

Cyclists, however, were not willing to give up the ghost. Whether through necessity or choice, many continued to cycle, driving complaints and lamentations from the increasingly powerful road lobby. Crucially, however, Amsterdam cyclists recognized the true nature of this fight—balancing the persistence of individual travel on the roads with an organized movement of political tenacity and cultural innovation. They didn't simply urge cyclists to keep cycling, but rather they focused their efforts and demanded a seat at the table in designing how the city would operate.

Tactics included non-violent resistance (slow bike rides), electoral politics (pro-bike parties began winning seats on the city council), and absurdist theater (protesters asked drivers to kill their engines and offered to push their cars instead). Some activists even encouraged more confrontational guerilla tactics, including unauthorized installation of bike racks and crosswalks, the supergluing of locks on illegally parked cars, and the clandestine scratching or denting of luxury vehicles.[4] The anarchist group Provo organized one of the world's first bike-share schemes, and when pedestrian deaths peaked in 1972, groups like *Stop De Kindermoord!* ("Stop the murder of children!") staged mass die-ins. In doing so, they applied an explicitly moral lens to the often dry world of city planning and transportation infrastructure.

Historical Serendipity

Chris Bruntlett is a Canadian author and cycling advocate who—like Pete Jordan—also moved to the Netherlands with his family after having witnessed the glory of Dutch cycling. (Together with his wife, Melissa Bruntlett, Chris co-authored a book called *Building the Cycling City: The Dutch Blueprint for Urban Vitality*.) Based in the western Dutch city of Delft, he now works as marketing and communications manager for the Dutch Cycling Embassy, a public-private partnership that helps communities around the world to learn from the Netherlands' experience with pro-bike policy and planning. When we connected via webcam to talk about Dutch cycling in general, and about Amsterdam in particular, Chris reiterated a point also made in Pete Jordan's book.

While the bike advocacy of the late '60s and early '70s was important, there was also a significant element of historical luck involved in turning the tide on cars. Specifically, both Chris and Pete Jordan point to the fact that the renaissance in Amsterdam's (and by extension the Netherlands') bike culture coincided with the oil crisis of 1973, which prompted a Sunday motoring ban and a growth in the sale of bicycles. Community groups organized parties and gatherings in spaces previously reserved for cars. TV reports even showed Dutch drivers humorously harnessing their cars like buggies to the back of horses. And, according to Chris, this temporary shift in power structures created space for alternative visions of what cities could and should look like:

> The streets and the motorways were opened up to the people. People were out there roller skating and horseback riding and having picnics. And they suddenly realized how much space they had given up to the private automobile. A lot of people point to that as a real lightbulb moment amongst the citizens that helped them understand where this trajectory was taking them.

Over time, the government began responding to the change in public opinion. Gradually, planning priorities were shifted back toward biking, pedestrianization, and livable streets. Critically, a significant effort by Dutch planners went not just into building out bike infrastructure, but also into making driving less pleasant or convenient for local transportation. Speed limits were slowed down and cars were deprioritized in planning decisions. Rights of way at intersections were renegotiated. (In the nearby city of Rotterdam, some traffic lights are now timed to give cyclists more priority when it's raining.)

On average, Chris explained, it now takes nearly eight times as long to drive the same distance by car in the center of Amsterdam as it does to ride a bike. In other words, there has been a collective and focused effort to redistribute power in the city—and individuals have responded accordingly.

As a result, nearly half the city's population commutes by bike on over 300 miles of dedicated bike paths.[5] And while many are undoubtedly motivated in part by climate or environmental concerns, it's fair to say that the majority simply do it because it's one of the easiest, cheapest, most enjoyable, and most culturally accepted forms of transportation within the city. In fact, Chris told me, Amsterdammers—and the Dutch in general—will often look confused when an enthusiastic foreigner expresses their admiration for Dutch "bike culture." And he largely sees this as a function of just how universal the bike has become as a generalized mode of transport:

> Most people who cycle now in the Netherlands have absolutely no idea of the blood sweat and tears that were spilled on their behalf to create the conditions they are currently enjoying. Yet almost everyone rides a bike here, whether they are five years old or ninety years old. It has no boundaries in terms of age or economic status or ethnicity or physical ability. In fact, one of the fastest growing groups of people riding bikes here are people with physical disabilities, who might ride an adapted tricycle or hand cycle or some other amazing machine. It's enabling people to lead amazing, autonomous lives. Yet they wouldn't consider themselves a cycling fanatic or enthusiast.

While Chris was right that the Netherlands has a growing number of options for disabled bikers, it is worth noting that disability activists have also highlighted accessibility challenges posed by Dutch bike infrastructure—especially for those who can't bike.

While challenges remain, there's a lot that the climate movement can learn from the Dutch bike wars. And it was increasingly clear from my conversations with Chris that the lesson is not that politics matter more than individual lifestyle change. Indeed, the efforts of activists would never have succeeded if individual citizens—from all walks of life and all stripes of politics—had not also chosen to take up the saddle. And that especially includes those who stubbornly persisted when it felt like they were swimming against the historical and cultural tide.

The lesson is, however, that individual action can only take us so far. Any successful attempt to change behaviors must engage with the systems that shape those behaviors. And it has to do so while pulling together a coalition that's broad enough to change the rules of engagement:

> It was a really broad coalition across The Netherlands and in Amsterdam in particular [that made change happen]. I think that's worth emphasizing. It wasn't just the left-wing anarchists. It was families who didn't want to see speeding traffic through their neighborhood. It was historical preservationists who didn't want to see the demolition of neighborhood character. It was public health institutions, and all of these other organizations, who found common ground under this banner of building a more livable city, that also happened to be a more sustainable city.

Amsterdammers came to see the bike as an integral part of the story of the city, and they re-embraced cycling as a way to write themselves back into that story. It was a story that played out in the civic realm—with people willing to protest, advocate, agitate, organize, and run for office, just as much as they were willing to get on their bikes and ride. And they were willing to do all that with a defined, focused vision for what they were trying to achieve. The result was change that spread well beyond a "coalition of the willing," and instead shaped the behavior of an entire city—a city which is now exporting its cycling expertise to the world.

This story, I believe, serves as a powerful allegory for what we need to achieve in almost all aspects of the climate fight. As famed climate scientist Michael E. Mann—whose 2021 book *The New Climate War* explores how fossil fuel industry uses individual action to distract us from systemic insterventons—once argued in an article for *Time* magazine, "We need systemic changes that will reduce everyone's carbon footprint, whether or not they care."[6]

The Real Power of the Individual

By now it is (hopefully!) abundantly clear that I am not opposed to individual action. As almost everyone I spoke to in this book emphasized, we need *both* systems-level activism *and* for people to start making changes in their own lives.

It's just that the changes we make in our own lives matter for entirely different reasons than we've been told.

We don't need more people to ride a bike because it will cut their personal carbon footprint. We need them to do so because it will send a signal to politicians, planners, businesses, and fellow citizens. That signal, along with organized activism—and support for that activism from folks who aren't yet ready to ride—will in turn help to change the systems that make cars the default choice in far too many situations.

Similarly, we don't need every single environmentalist to go vegan because eating meat is an eco-sin and our movement demands purity. Yet we should embrace and applaud vegans (and reducetarians, and weekday vegetarians), because their dietary choices are disrupting the industrial food complex and creating a culture where all people are more likely to eat plants. So here's where the conundrum gets interesting: If the behaviors the climate movement is promoting are often the same, for example eating less meat or driving fewer cars, does it really matter whether the case we are making is about personal virtue, or more systems-level intervention?

My answer to that question would be a resounding "Hell yes." And it's a lesson that applies well beyond the climate crisis too.

A Reckoning on Race

On May 25, 2020, just as many states were beginning to ease lockdown measures related to the COVID-19 pandemic, George Floyd—a 46-year-old Black man—was murdered in broad daylight by Minneapolis police.

George Floyd was far from the first nor the last Black person to die at the hands of the police, of course. Just weeks earlier, Breonna Taylor—an ER technician—was shot to death in her own bed by officers of the Louisville Metro Police Department. Officers even stepped over her body as she lay dying, failing to administer any form of aid or comfort. These deaths, coming on top of America's centuries-old history of racial injustice, ignited an uprising of furious intensity.

This uprising manifested in many different ways. Protesters took to the streets, statues were toppled, and even the cautious and conservative world of corporate America stepped up and declared that Black Lives Matter. We also saw another, more reflective side of the discussion, as white people began interrogating their complicity within the system. Sales of books like *White Fragility* shot through the roof, and countless columns were written and conversations were had on privilege, white guilt, and allyship.

Undoubtedly, the renewed focus on our legacy of racism—and on the complicity of individuals who have benefited from that legacy in one way or another—prompted many to take worthwhile and meaningful action. There were also some commentators, however, who worried that concepts like privilege and guilt might serve as a distraction from the real work of challenging and transforming the system. This sentiment is not exactly new.

It's Not About Me (Or You)

Back in 2019, writing for a progressive online Christian publication called *The Salve*, Ajah Hales challenged the utility of white guilt. In doing so, she offered a provocative analogy in which she asked us, as readers, to imagine we've come across a badly injured assault victim. As we spring into action to help them, we discover we've left our cell phone at home and, to make matters worse, we also remember that we never did take that CPR class we'd been planning to enroll in.

What, she asked, should our next step be?

> Perhaps you would run to the nearest store or house and ask to use their phone. Maybe you would check to make sure the person is still breathing. Maybe you would check his/her pockets for a phone.
>
> How much time would you spend pacing beside the person as they lay dying, berating yourself for not having your phone and never taking a CPR certification?
>
> Probably none, right? Because this is a life or death situation; it's not about you, and your guilt is worthless in this scenario.[7]

She went on to specifically address her white "siblings in Christ" with a particularly pointed message: "Your white guilt isn't helping me."

Systemic racism is, by definition, about the system. That doesn't mean we get to ignore our moral duty. But it does mean we have to prioritize interventions that move beyond the realm of individual agency. In other words, we have to accept the fact that systemic solutions for all are more important than a personal sense of absolution for some.

We can't fix gentrification or segregation simply by choosing where we want to live, or how much we sell our homes for. In fact, it would be madness to center the discussion there. Instead, we need to dive deep into tax laws and planning, school policies and community dynamics, law enforcement and social policy. And then we need to figure out where we, as individuals, most effectively plug in to the ongoing quest for justice.

Unfortunately, grappling with systemic issues is difficult.

The Lure of Agency

As a teenager, I once took the two-hour train ride from Bristol, in South West England, to London. A group of activists were gathering to Reclaim the Streets. They staged a minor car accident outside Angel Tube Station with two sacrificial, junkyard vehicles. They then used the chaos that ensued to start an impromptu, illegal rave in protest at car culture. Hippies danced. Music blared. And, in one small corner of the city, London's never-ending flow of traffic came to a standstill as we explored alternatives.

A few years later, a similar action shut down the M41 motorway in central London. News cameras showed footage of a troop of stilt walkers dancing in front of a line of riot police , while loud techno blared from a portable sound system nearby. The police only later realized that the skirts of those stilt walkers, and the pounding music, had been concealing an army of jackhammer wielding activists. When the stilt walkers retreated, they revealed a surreal scene in which the asphalt of the motorway had been removed in order to plant a temporary, fledgling forest.

As a young idealist, I was enamored by the audacity of direct action protest. And I was sympathetic to the idea that civil disobedience—and perhaps even damage to property—was justified in the face of the destructive efforts of the fossil fuel industries.

But I was also skeptical. For all the "no more roads" rhetoric, these protests were alive with the sound of chugging diesel generators and clapped-out old hippy buses. How could we demand change from others, if we weren't willing to walk the walk ourselves? Perhaps more importantly, could we really expect a rag-tag group of party goers to actually dismantle the complex economic and cultural structures that had made fossil fuels so dominant?

At the same time as I was learning how to protest, I was also introduced to another, gentler side of the counterculture. I learned how to compost. I started working at a vegetarian restaurant. And I regularly joined a group of activists to plant native woodlands above the denuded Calder Valley in northern England—a community that was already feeling the ravages of climate-driven weather and which has continued to be hit by devastating floods in the decades since.

While the protests and illegal street raves had been thrilling, I was profoundly relieved to find ways to put my own house in order. Part of that relief came simply from walking the walk—meaning there was less friction between the ethics I espoused politically and the life I lived personally. But I recognize now that there was another source of relief at work: the fact that it just seemed easier to address what I had direct control over, rather than to engage with the painfully slow process of societal-level change.

I didn't have *leverage* to overturn the system, so I found myself retreating to where at least it felt like I had some *agency*.

How Change Actually Happens

Austin Choi-Fitzpatrick is an author, speaker, and educator who has spent a long time thinking about how social change actually happens. Introduced to me by Minh Dang—a mutual friend and a former professional collaborator—Austin was billed as being smart, busy, and having an exuberantly rocking beard. Minh was not wrong on any of these counts.

Describing himself as a structuralist, Austin told me from the get-go that he believed there are much bigger forces at play than our own individual lifestyle choices. Expanding on that idea, he explained that social change is the product of a powerful and complex interplay of forces, and that these range from specific economic and technological changes to gut-level popular reactions to whoever is in power at the time.

When we managed to connect over Zoom in the summer of 2020, I asked Austin why he thought that so many climate-concerned individuals—particularly in the West—were often drawn toward taking action on a personal level. He politely suggested that I might consider asking a slightly different question:

> America's ethos—plus consumerism—breeds narcissism. We're just high-strain narcissists. But who cares if people have an over-developed sense of agency? It might be a lie, but let's not even worry about that. Rather than telling them it's a lie, how can we instead accelerate it and point it in the right direction?

As our discussion progressed, I referred Austin to the idea—discussed in Chapter 9—that vegans get criticized not because people disagree with their ethics, but rather because many of us who agree with them can't quite summon the same level of commitment. I pondered whether there was some way for animal rights or environmental food activists to neutralize that reaction, and his eyes lit up.

Rather than try to neutralize it or worry about it at all, he argued, activists would do well to view it as a sign of pent-up demand for their ideas. Instead of spending time trying to convince guilty meat eaters to go 100 percent vegan, we might be better placed trying to change the context within which they are making their dietary choices:

> Don't try to convince people to be vegan, make a goddamn veggie burger that tastes like blood. So if you're thinking about what people should do, one of the most important things they can do is to change the structures that shape their behaviors.

Austin's assertion—that it was better to focus on changing structures, rather than on changing our own behaviors within those structures—got to the heart of why I had started writing this book in the first place. It also summed up a theme that had become apparent throughout the conversations this project had led me to.

It's not that we don't need people to reduce their meat intake, ride a bike, or limit their flying. It's just that simply asking, encouraging, or even berating them to do so is not necessarily going to work. Given that our behaviors are shaped so strongly by the environment we live in, we should be focusing the majority of our efforts on where we have the greatest opportunity to shape that environment for all of those around us.

Not only did this perspective point to a more effective way to do change, but it also—I surmised—offered a way out of some of the guilt and self-blame that many climate-aware individuals grapple with.

What's My Duty?

I was thinking about this idea when I reached out to Julia K. Steinberger. At the time, Julia was a professor of ecological economics at the University of Leeds in Yorkshire, England, as well as a contributor to the Intergovernmental Panel on Climate Change (IPCC) 6th Assessment Report. (She has since moved to the University of Lausanne, Switzerland.)

I first came across her work when she was interviewed by Jocelyn Timperley for an article about guilt, blame, and the climate crisis for the BBC. In that interview, she too suggested that we tend to focus too much on the immediate impacts of each individual's actions, as opposed to the systems that bring those actions into being:

> "Just because you can allocate [emissions] to an entity or to a location in a supply chain, does not mean that the power of agency lies with that entity or that location in the supply chain," says Steinberger. "If you're thinking about these supply chains, are you going to say that final consumers actually have the final decision-making over everything that happens upstream? Who is actually taking the damaging decision?"[8]

Describing herself as always having been an activist, but a disillusioned one, Julia explained to me that coming to grips with the enormity of the climate crisis—compounded by the election of Donald Trump in 2016—led her to become a different, more vocal type of academic than academia had taught her to be. As part of that awakening, she came to focus very intently on the question of what we are called to do:

> The roles we've been given—handed down from the Enlightenment or the Industrial Revolution or whatever our cultures came up with—are not suited to our time. That's when I decided to do things differently, and to think about the fact that we are little creatures who have to interact with

these big monsters that are using us to destroy the world....
We need to learn over and over again how we can become
forces for revolutionary action, because that's never been
more desperately needed.

This idea that we didn't choose the circumstances of our existence,
but that we have a responsibility to engage with them, is central to
understanding if and how we can behave as ethical human beings
during the climate crisis. Julia was careful, for example, to differ-
entiate between the idea that personal guilt or shame is not necessarily justified or helpful, versus the broader (and more relevant)
concept of what our responsibility is to act:

One of the things I try to tell people is that maybe it's not
their fault—but once you become aware of what the system
is, and what the system is doing, you don't actually get to
go back and be comfortable. I don't care about morality or
purity or whether you feel good about yourself or not. This
is about something that is so much bigger.... You get a few
seconds to sit and have a beer and feel sorry for yourself that
you were born into these times and you have to make these
tough choices, and then you really need to get over yourself.
You need to get in the fight.... If you're worrying too much
about your own feelings, then it's self-indulgence and it's a
waste of time.

Julia's suggestion—that even worrying about your personal guilt or
culpability was ultimately little more than navel gazing—appeared
to be a somewhat more blunt take on my own concerns about an
excessive focus on "lifestyle activism." It wasn't that the actions
being proposed were necessarily wrong. It was just that they centered our focus on the wrong unit of measurement.

In fact, while it might be tempting to interpret Julia's counsel as
a call to focus only on overtly political efforts, she has by no means
rejected the value of personal action. She has, for example, stopped
flying. She has urged people, especially the globally affluent, to

curb excessive consumption. She advocates moving your pensions and other investments away from fossil fuels and other polluting industries. And—until she had found herself homebound due to a sporting injury—she told me she had been a determined and visible cyclist, including when she was seven months pregnant.

Citing the work of Peter Kalmus, who I interviewed in Chapter 8, she argued that one of the most important—and sometimes difficult—things we can do is to be public and visible in the efforts we are making. She suggested that crossing that threshold into overt climate activism can make it significantly easier to keep going down that path:

> Once you do one thing, then taking those other steps isn't that hard. It's harder to do things publicly. But once you've decided to change how you are interacting with the outside world to both lessen your own impact, and also to try to shift other people to see things differently,...that first step of doing it in a public way makes things significantly easier. I didn't fly for work for something like six years before I told people I wasn't doing it.

Ultimately, she said, it was less important to argue about where folks get started on their journey to climate action, and instead make sure that they keep moving down the path toward creating real, large-scale, society-wide change. But what if the people you're surrounded by aren't ready to talk about the climate crisis at all?

Shifting Our Collective Values

Elizabeth Koebele's career has put her in a somewhat strange place. Despite her PhD in environmental studies, she eventually became assistant professor of political science at the University of Nevada, Reno. As such, she now finds herself spending much of her time teaching classes where half of the students have never studied climate or environmental science, and where the other half have a solid understanding of the science but have never really had the opportunity to think about it within the political realm.

This intersection provides Elizabeth with an interesting vantage point for looking at where the hard realities of science meet the somewhat more intangible forces of culture and politics. Her work has often focused on the topic of water resource management:

> I have this background growing up in the desert and seeing scarce water resources. Many of the people who live in that environment are seeing the impacts of climate change in their daily lives. They see snow packs melting earlier, changes in stream runoff, and yet they don't want to talk about climate change.
>
> So water resources became this way to talk about climate change without having to say "climate" all the time, and using our shared resources as a place where we may be able to gain more political consensus.

As we talked, I reflected on the idea of conflict around water as a useful parallel to discussions around climate, carbon emissions, and individual versus societal-level solutions. After all, water is inherently and undeniably a collective issue. You can take as short a shower as you want, but if a neighboring home or farm decides to suck up all the groundwater, your own conservation efforts are going to mean little to nothing—unless you can leverage your own actions to influence the behavior of those around you.

Elizabeth has some experience of seeking that leverage, having studied and developed processes at the local, regional, and even international scale that bring people together. The key, she says, is often to find ways to encourage or enhance collaboration to protect shared resources:

If we can get farmers and environmentalists together to protect water quality, then environmentalists like that for habitat and recreation, and farmers like it for growing high-value crops in different places. These processes take a lot of time, but they might be a realm where we can find points of leverage to create change beyond our own individual action.

There are challenges, of course, to a collaborative approach. In addition to the sheer time and effort it takes to build consensus, Elizabeth told me, it can also lead to fairly middle-of-the-road outcomes where nobody really gets everything they want. Still, there may be clues here about how we can forge similar breakthroughs on climate. And while Elizabeth has sometimes railed against the centrality of individual behavior change too, she suggested that it may have a role to play in helping people to visualize what comes next:

> It's a conversation that a lot of my friends have about their lives. I'm a cyclist, I bike commute, I'm a vegetarian. And a lot of my friends have similar lifestyles—and yet we are all asking what else we can do, besides not living.... While I often encourage people to think more structurally, we can still use those individual actions to set an example, motivate people, and get people to coalesce around bigger political movements.

To put it another way, if the climate movement is going to continue to talk about individual action in our personal lives, then we need to get better at communicating its true value, and at encouraging people to see their individual lifestyle changes through a collective lens of mass mobilization. We aren't on a quest to find the perfect green lifestyle for ourselves. We're on a quest to identify targeted interventions that—when combined with millions and millions of other interventions—will eventually transform the entire system.

11

Climate Hypocrites Unite!

I moved from the UK to the United States in December of 2006.

As I shared earlier in the book, it was an exciting time in my life. I was 28 years old, in love, and soon to be married. Jenni and I began exploring the possibility of buying a house, starting a family, and solidifying our career paths.

Many of the life decisions we made back then involved some effort to lessen our carbon footprint. We started raising chickens. I (unsuccessfully) tried my hand at beekeeping. I telecommuted from a shed in our backyard. I even began brewing my own biodiesel made from waste vegetable oil. And Jenni and I took the leap of installing solar hot water on our new home.

Here's a somewhat sobering fact, though.

For all of my efforts to tweak my carbon footprint, it actually increased by an order of magnitude during those early married years. Part of that increase came from the trappings of middle-class, conventional adulthood. But most of it was simply because I moved from one country—where renewables and nuclear were more commonplace—to another, where more coal was burned, where fridges and other appliances were larger and less energy efficient, and where it was harder to live without a car or take public transit.

Meanwhile, the majority of my old friends in the UK saw their own carbon footprints slashed. Regardless of their interest in climate or the environment, their eco-scorecard improved as coal use plunged and renewables—most notably offshore wind and rooftop solar—took off at a rapid pace. By 2016, the UK's annual carbon emissions had fallen some 36 percent below the highs of the 1990s. The last time Britain's emissions had consistently been this low, Queen Victoria had been sitting on the throne. And while some of the early drop off in emissions could be attributed to heavy industry and manufacturing moving overseas (a.k.a. offshoring emissions), according to an analysis by climate news organization Carbon Brief, the UK's consumption-based emissions—a calculation that includes the impact of goods and services brought from overseas—also fell at a dramatic pace from 2006 onwards.[1] While this elevated speed of emissions reductions was still nowhere near enough to meet the goals set out in the Paris Agreement, it was clear to even the most casual observer that Britain was making considerably more progress than my new adopted homeland of the US.

That's the thing about societal-level challenges: They need to be tackled at the societal level, and doing so requires us to understand that there are powerful actors involved who would prefer to maintain the status quo.

A False Dawn

While it seems like a distant memory now, the early part of the 2000s had actually been marked by a similar sense of bipartisan progress in the US too. At the state level, legislators around the country were enacting net metering laws and/or renewable energy mandates, and solar and wind were booming as a result. Even relatively conservative states like Iowa and Texas suddenly saw an influx of investment and well-paying jobs in the renewable energy field. In fact, many advocacy groups believed we were on the cusp of a dramatic shift in our energy mix that might have rivaled the rapid progress being made on the other side of the pond.

Unfortunately, this was not to be.

Fearing a threat to their business model, a coalition of energy companies, industry groups, and aligned politicians hatched a plan, and then quickly got to work undermining this emerging new paradigm. As detailed in Leah Stokes' excellent book *Short Circuiting Policy*, their efforts didn't simply seek to shore up incumbent industries. They also stoked partisan division with a view to boosting fossil fuels and consolidating their power:

> At the same time that these interest groups lobbied for retrenchment and repeal, public opinion and legislators' positions on clean energy policy became increasingly polarized. Through lobbying politicians and regulators, and driving polarization in the parties, the public, and the courts, these opponents often succeeded in weakening clean energy laws.[2]

It goes without saying that such obstructionism was morally reprehensible. In some cases, it was probably illegal too.

Just as I was finishing the first draft of this book, the political world of the Midwest was rocked by news of an FBI raid. Ohio House Speaker Larry Householder, a Republican, was taken into custody from his farm by federal agents on charges of racketeering. Specifically, the Feds alleged that Householder had been involved in a $60 million bribery scheme linked to Ohio's controversial House Bill 6—a piece of legislation which gutted the state's renewable energy and energy efficiency subsidies. While Householder has yet to be convicted at the time this goes to press, two of his associates have already pleaded guilty.

In other words, even while I was raising chickens, obsessively switching off lights, and schlepping used cooking oil to brew biodiesel in a shed with my friends, I was only ever going to tinker around the edges of my environmental impact. Between the structural underpinnings of the American economy and the active (and apparently sometimes criminal) efforts of Big Energy, it was almost inevitable that my footprint would now be bigger as a "greedy American" than it ever had been as a Brit.

The Power of Imperfection

Recognizing that our behaviors are shaped by the structures we inhabit isn't about passing the buck. And accepting our imperfections isn't about assuaging our guilt or giving ourselves excuses to keep on doing as we please.

It's simply about becoming more effective as activists.

By embracing our supposed hypocrisies, we undermine the arguments of our opponents—those who would use our shortcomings as an excuse for doing nothing. More importantly, in doing so, we are also able to get a better handle on exactly what the path to society-wide decarbonization might look like. The point is not to ignore what we can and should be doing in our personal lives. Instead, it's to see those actions within the true context that gives them both meaning and power—and that's in how they help shape the world around us.

Finnish social innovation experts Jussi Impiö, Satu Lähteenoja, and Annina Orasmaa, of the think tank Demos, have coined the term "carbon handprint" to help conceptualize the idea of measuring not just our negative impact, but also the positive influence that each of us has in moving toward society-wide decarbonization.[3] It's a powerful idea that's worth repeating: rather than just measuring our success in terms of being less bad, we can start to think about our impact in more positive terms.

Here are some of the ways that I've tried to do that in my life, and what a similar approach might look like for yours (I have included a list of recommended resources, organizations, and actions at the end of the book for those wanting more).

Lead From Where You Stand

My suggestion for anyone looking to make a difference is to start not with a generic top ten list of How to Cut Your Carbon. It's not even to focus on your own carbon footprint at all. Instead, take a frank and honest look at where you have the greatest opportunity to create wider-scale change:

- What does your social network look like, and how can you have an influence over those you love?
- What issues, organizations, or activist groups are you drawn to, and how could you get more involved?
- What opportunities do you have at work, at school, or in your community?

- What power, privilege, or advantages do you enjoy that you could leverage in service of the movement?
- What strengths, skills, or knowledge do you bring to the climate fight?
- What do you love doing? What sustains and motivates you over the long term?
- And, crucially, what forces stand in your way, and what needs to happen for those forces to go away?

You'll have plenty of time to learn about how you currently cause harm. Yet I believe you should start with an expansive and ambitious view of where your unique power lies.

Yes, Your Footprint Does Still Matter

Once you have that understanding of where you find yourself in the world, there is indeed value in understanding the nature of your carbon footprint. Not only will it allow you to identify some specific ways that you can meaningfully reduce it, but it may also help you get a better understanding of where the system gets in your way.

When I calculated my carbon footprint for the first time, for example, I was shocked by how much my work-related travel contributed. This realization led me to have conversations that changed my employer's travel policies.

The trick is to think about a low-carbon footprint not as an end goal in itself—after all, your carbon footprint is infinitesimally small when looked at in isolation. Instead, the calculation becomes a useful metric for identifying which behavior changes are significant enough to really put pressure on the wider system, and which behavior changes are onerously hard or unattractive and therefore may require a systems-level intervention.

Think you can give up flying? Go ahead and do it. But even if you can't, consider cutting back on the number of flights you take or—if you happen to be privileged enough for this to be relevant—switching even some of your journeys from business class to economy. Think you can live without a car? All power to you. But even if you can't, find ways to use your car less, or switch to a greener model. Remember, your goal is not to get your own footprint to zero so you can feel better about yourself. It's to do what you can to put pressure on the system, and in doing so to reach tipping points that ignite change, eventually making it easier for everyone to do the right thing.

Whatever you do though, be careful not to take it too far.

Focus on Your Net Impact—And Make It Positive

For a brief moment in my early twenties, I found myself living alone in a tiny, practically windowless apartment on the outskirts of Bristol, England. Already worried about the impact of climate change, I started a rather depressing personal habit: once darkness fell, I would turn out the lights and spend the evening listening to my hand-crank solar radio in the dark.

Meg Ruttan Walker—the Canadian activist I spoke to about baby shaming in Chapter 5—has described extreme efforts to eliminate our own footprint as being somewhat reminiscent of an eating disorder, a condition with which she herself has struggled. We can become so obsessed with our own complicity, or our disgust at the state of the world, that we focus almost exclusively on purifying *ourselves* of the ills we see around us, rather than actually fixing those ills on a societal level. Such extreme asceticism is difficult to sustain for most of us. Instead, I personally have chosen to measure my progress based on three simple questions:

- Firstly, where can I have the biggest positive impact on the climate fight?
- Secondly, how and where can I meaningfully reduce my carbon footprint within the societal context I inhabit?
- And thirdly, on balance, am I doing more good than harm in moving our society forward?

We will eventually all have to find our way down to zero emissions within the next couple of decades. But decarbonization is a society-wide undertaking. It's not incumbent on any one of us to get too far out in front.

Double Down Before Diversifying

Because the climate crisis is all encompassing, it can quickly become overwhelming. Instead of trying to tackle all aspects of your own lifestyle at once, I recommend first looking to deepen the impact of whatever part of the fight you have chosen to focus on.

- **Passionate about composting?** Look for ways you can take that passion and spread your influence to others—by starting a composting program at your kids' school, for example, or teaching co-workers how to make a worm bin.
- **Started biking to work?** Look for other cyclists you can connect with, join a bike advocacy group, or engage with your elected representatives about cycling policy.
- **Cutting back on conference travel?** Talk to conference organizers about developing virtual alternatives, create tips and resources for co-workers who would like to do the same, and work with your employer or professional affinity groups to create permission structures for everyone to travel less.

Above all, remember that our behaviors and our decisions are governed by structures and influences that are bigger than any one of us can control. That's why it's important to calibrate your efforts so that they put pressure on those structures and help pave the way for others to follow. (And don't forget to get out there and vote!)

Connect the Dots

Even as we specialize in our own particular area of focus, we can and must learn to connect the dots with what others are doing too.

More focused on cutting your own carbon footprint? Great—but try to show up to a protest or two and support those who are fighting Big Oil. More interested in direct action or community organizing? Chances are you can find a little time in your life to also set up a composting system, or use your car less. The point in concentrating your effort into one or two areas of intense focus is not to suggest that other areas do not matter. Instead, it's to start seeing yourself as one part of a much bigger movement, and to trust that others are working on this too.

Be Kind to Yourself

When Naomi Klein addressed the students at the College of the Atlantic during their commencement address in 2015 (referenced in Chapter 2), she took a moment to explore one of the thorniest challenges of the climate crisis:

> One of the real dangers of being brilliant, sensitive young people who hear the climate clock ticking loudly is the danger of taking on too much. Which is another manifestation of that inflated sense of our own importance.[4]

As an example, she referenced a recent phone call she had had with a young protester who, during that very call, was suspended from a bridge, blockading an oncoming oil drilling vessel. While most of us would look up to that protester as a hero, this young woman was—according to Klein—in tears. She just wasn't sure if she was really doing enough. While the example may be extreme, the principle behind it is widely and deeply felt: "Am I ever doing enough?"

On the one hand, this is a valuable question. This is, after all, an emergency. Yet unlike a house fire—where the decisions we make might literally mean *immediate* life or death for ourselves and those around us—we must recognize what is, and what is not, in our power. The climate crisis is going to last a very, very long time. We're going to be living in it for the rest of our lives. I strongly recommend picking your battles carefully.

Don't Judge (Except for the Real Assholes)

If we're going to be kind to ourselves, then we need to extend that courtesy to others too.

As we've already seen, there is much power to be had in choosing not to fly—and in teaming up with others to encourage wiser travel choices. Yet I believe we need to be careful about blanket pronouncements or divisive attempts at claiming the higher ground. Likewise, the act of going vegan or living car-free can be liberating and incredibly meaningful. Yet we need to recognize that berating others is unlikely to get them on our side. There is just no way we can build a movement on the scale that is needed if we are going to spend our time policing each other for perceived, personal transgressions that are largely accepted and even actively encouraged by the wider world around us. Instead, channel your energies into understanding the barriers that prevent others from acting—and then fight like hell to abolish them. And yes, that does include shaming the hell out of the powerful entities that have profited from business as usual.

Remember, this isn't simply a moral calculation about who is actually "guilty." It is primarily a tactical decision about deploying judgment where it will be most effective. As Jennifer Jacquet explained to me in Chapter 5, shame is a limited resource. So remember to use it wisely.

Finding Our Place

The ideas laid out in this book are by no means an exhaustive list of principles, nor are they necessarily the only way to start thinking about your carbon handprint. They are, however, my attempt to codify something I've spent an awful lot of time worrying about: How can each of us help to enact solutions, even as we recognize that we are integrally a part of the problem?

In exploring this question, I've been hampered by another, equally thorny challenge. And that's the fact that I am by no means certain what a society-wide solution to the climate crisis will actually look like. In fact, I am increasingly convinced that nobody—meaning even the smartest and most qualified people in our movement—knows exactly what it will take to get us where we need to go.

What we do know is that humanity can and must dramatically reduce its collective carbon footprint, we must do it fast, and we must also do so while empowering lower-income nations and communities—the folks who have contributed the least to the problem and stand to lose the most—to raise their standard of living, build their resilience, and free themselves of the exploitation and oppression that has too often held them down. In the end, we're all going to have to find ways to enact society-wide change that includes at least the following:

- defunding and disempowering the powerful corporations that profit from destruction
- creating community-led solutions that address not only climate change and environmental pollution, but do so through a lens of social and environmental justice
- radically reducing the cost of doing the right thing—and developing low-, no- and negative-carbon alternatives for transport, energy, manufacturing, housing, and more
- protecting, regenerating, and restoring the natural systems upon which all true prosperity is based
- developing genuinely useful economic models that prioritize human well-being, happiness, inclusion, and learning over material gain and excessive private wealth

In order to achieve these goals, it's never going to be enough to live an ecologically pure lifestyle in a world that carries on as if everything is normal. But we also can't ignore the ways that our own choices reinforce and sustain the system. Writing in a series of Tweets, sociology professor Daniel Aldana Cohen described the challenge like this:

> To address the climate emergency, we need non-linear political economic change. But non-linear change doesn't come from nowhere; it requires cumulative incremental change. [The] question isn't incremental vs. non-linear. It's *which* incremental strategies best lead to non-linear.
>
> [Because] cumulative incremental change is necessary, but not sufficient. But since we face constant decisions on incremental matters, we should make them intentionally.

> Non-linear change involves structural ruptures we often can't control. So we do our best [with] what we can control.[5]

Make no mistake: These are monumental tasks and the stakes could not possibly be higher. We are going to need every single ally we can get. And that's why we need to move beyond the accusations of hypocrisy that have too often caused a split between those who should otherwise be aligned. Each of us is an important, imperfect character in a story that is much larger, more complex, and multifaceted than any of us can even imagine. Once we acknowledge that, we can start moving forward together.

Most of us are climate hypocrites in one way or another. We urgently need to find a way to make this work.

Coda:

The Journey Down, Together

Picture this: You're hiking high up on a rugged mountain.

With you is a diverse group of fellow travelers, each with different physical abilities, strengths, weaknesses, viewpoints, and personalities. A few of your party have been lagging behind on the expedition due to persistent injuries. Others just don't seem to be motivated or able to move too fast. Yet others are hampered by crappy equipment, because the people organizing the expedition weren't exactly equitable in how they doled out provisions.

As you make your way along a particularly precarious ridge, one of your fellow hikers holds up a hand and urges the rest of you to be quiet. Her radio is on, and there's an urgent warning being relayed. A life-threatening storm is just hours away, and you need to get off the mountain now.

Making matters more complicated, you know that at the end of your hike there is a deep, treacherous swamp. The only way to cross that swamp is to form a human chain—a human chain for which you will need every single one of your current party to be present.

In other words, leaving someone behind is not an option, even if you wanted to.

Soon, a plan starts to formulate. A small group is designated to plough ahead, serving as an advance party. Their role is to identify the best route down, and to lay the groundwork by clearing brush, tying ropes, and removing obstacles from the path.

The rest stay behind. And you divide up tasks among you:
- some focus on packing up camp as efficiently as possible for transport
- a small group gets together to ensure you have enough provisions to keep you fueled as you descend, and that they are distributed fairly to those who need them
- a third cohort focuses on tending to the injured and on figuring out a safe way to transport them
- another begins to follow the advance party—communicating between the groups and laying a specific set of markers for those who follow

Together, you slowly start to make your way down the mountain. And you—as an individual—have to find your most effective place within that group. Assuming you're in one of the lead groups, you're faced with a unique challenge. Get too far out in front and you've lost your opportunity to lead. But stick with the main group and there's a danger the pace won't be quite enough to save you all.

So you modulate, you adapt, you coordinate, and you communicate—not without argument or disagreement—but with a broad understanding that the success of your mission rests on finding a way to make this work. For everyone.

Eventually, however imperfectly, you make your way down. You take a moment to catch your breath, and then you survey the swamp that stands before you. You link arms. One of your party remains firmly rooted on solid ground, anchoring themselves to a tree.

The rest of you start tentatively making your way into the brackish and unpredictable water...

What Next? Resources, Organizations, and Actions

When I first started writing this book, my goal was to debunk the idea that individual action was a worthwhile focus for the climate movement. I couldn't have been more wrong. What I have come to believe, though, is that we need to learn to think about our actions through a much more systemic lens.

Below are some places to start.

It's important to note, however, that approaching the question of what you, specifically, should be doing to promote climate action will depend very much on who you, specifically, are. There are no one-size-fits-all solutions, and what might be productive or rewarding for one person might be largely futile or unappealing to another.

So take these suggestions as a starting point. Make them your own. And above all, get started on your own journey. Just remember to keep your eye on the bigger prize.

Knowledge Is Power

As promised, I have not spent too much time explaining the science behind the climate crisis, nor the solutions either. That's not because this knowledge is unimportant. It's just that other people, often much smarter than myself, have already done it. Here are a few of the books that have helped me understand the current state of play and to begin thinking about what we can actually do about

it. Getting a grounding in some of these books—and then thinking about how you fit into the crisis—will greatly help you identify
where you should be spending your energy:

- *Drawdown: The Most Comprehensive Plan Ever Proposed to Re-*
  verse Global Warming, Paul Hawken (ed.), Penguin Books, 2017.
  Edited by veteran environmentalist Paul Hawken, this book
  outlines 100 societal-level solutions to the climate crisis, and
  ranks them in terms of their potential impact. Ranging from
  conventional climate solutions like energy efficiency, mass
  transit, and solar power, to more surprising ones like the power
  of investing in women's education, the goal is to outline a comprehensive roadmap to reaching Net Zero, and eventually negative emissions, for everyone.
- *What We're Fighting for Now is Each Other: Dispatches from the*
  Frontlines of Climate Justice, Wen Stephenson, Beacon Press,
  2016. Written by a climate activist and journalist, this book offers a firsthand look at the emerging climate justice movement,
  exploring how a new and diverse generation of activists is moving past lifestyle environmentalism and instead taking on the
  might of the fossil fuel industries, as well as the systems that
  are designed to perpetuate destruction.
- *Youth to Power: Your Voice and How to Use It*, Jamie Margolin
  (Foreword by Greta Thunberg), Hachette Go, 2020. Written by
  teen climate justice activist Jamie Margolin—who I interviewed
  in Chapter 4—and featuring interviews with other young activists, this book offers a how-to guide for young people wanting
  to step up and make a difference. (There's plenty of wisdom
  here for those of us would-be activists who are a little longer in
  tooth too.)
- *Being The Change: Live Well and Spark a Climate Revolution*, Peter
  Kalmus, New Society Publishers, 2017. As a NASA climate scientist, Peter Kalmus—who I interviewed in Chapter 8—offers
  an interesting perspective for anyone wanting to dive deep into
  lower-carbon living, without losing sight of the fact that the

ultimate goal is not promoting voluntary abstinence, but rather society-wide decarbonization.

- *The Future Earth: A Radical Vision of What's Possible in the Age of Warming*, Eric Holthaus, HarperOne, 2020. Written by someone described as "the rebel nerd of meteorology" by *Rolling Stone*, this book takes on a rather remarkable challenge: articulating what our journey to a negative emissions society might look like over the next three decades. At once hopeful in its vision, yet unflinching in its assessment of our current predicament, it serves as both an alarming wake-up call and an encouraging blueprint for what we could eventually achieve if we get our act together fast.

Get Organized

Knowledge means little unless it is turned into action. And the single biggest thing you can do with your knowledge is to become an advocate for structural changes that make lower-carbon living easier. That doesn't mean you necessarily have to chain yourself to a bulldozer—although there are very good reasons to do exactly that. You could also choose to create change in your workplace or school, run for political office, or team up with neighbors to promote solutions at the local level. The point is simply to think beyond your own personal footprint, and instead seek collective actions that ripple outward across society. Below are some organizations you could consider joining, donating to, or otherwise supporting. The list is inevitably US-centric, but there are organizations in countries all over the globe working on different aspects of the climate emergency. So use these as an example and seek out people who are doing work that you are drawn to.

- **Sunrise Movement:** Youth-led and explicitly political, the Sunrise Movement is a relative newcomer within the climate movement, and yet it has already helped transform the debate. Its vision for a Green New Deal is ambitious, actionable, and aggressively focused not just on cutting carbon, but addressing

structural challenges like systemic racism and economic inequality too. (SunriseMovement.org)

- **Zero Hour:** Another youth-led organization with a strong focus
  on diversity and racial justice, Zero Hour was a driving force
  behind the climate strikes and marches that brought millions
  of young people onto the streets in 2019. (ThisIsZeroHour.org)
- **350.org:** Named for the fact that 350 parts per million is the
  level of carbon dioxide concentration that scientists consider
  safe for the Earth's atmosphere—a concentration that was exceeded back in 1988—this international organization is focused
  on moving humanity beyond fossil fuels. (350.org)
- **Dogwood Alliance:** I'm including this one because, as a board
  member, I am biased. Regional in its focus, and yet global in
  its impact, Dogwood Alliance has already transformed how
  many of the forests in the southeast United States are managed, and it is currently taking on the destruction of forests
  for supposedly renewable biomass energy. Dogwood has also
  been on the forefront of reimagining the climate and environmental movements through an explicitly racial justice lens.
  (DogwoodAlliance.org)
- **B Lab:** Created as the non-profit that coordinates the certification of B Corporations ("B" stands for Benefit), B Lab is also a
  terrific resource for any company that wants to lessen its impact on the planet and/or make its voice heard for climate action. (BCorporation.net)

Rethink Your Mobility

If you're going to take a look at your carbon footprint, then there's a
good chance that one of the largest pieces of your emissions pie is
how you move around. So seeking to find ways that you can embed
lower-carbon transport—or less travel altogether—into your lifestyle is a great way to make a difference, especially if you can bring
others along for the—ahem—ride.

- **Get a Bike:** If you are physically able to use one, getting a bike is
  a great first step to a healthier, happier way of traveling around

town. Given our focus on practicality, consider your machine carefully—e-bikes, cargo bikes, or simple, sturdy town bikes with a solid luggage rack are often a great option for those of us who are really just interested in getting from A to B. And given the fact that most of us live in less-than-perfect bike communities, don't forget to seek out bike advocacy groups who can help make your community safe.

- **Consider Your Car:** Getting rid of a car can be a great way to save money, reduce emissions, and stick it to the oil majors. But it's not practical for everyone. If for whatever reason you're unable to go car-free, electric cars are considerably less polluting than their gas equivalent; and they are getting cheaper, more ubiquitous, and often surprisingly affordable on the second-hand market. Car-sharing services also offer an alternative option—meaning you have access to a vehicle when you need one, but you don't have to pay for it while it's sitting in your driveway unused. Think about ways you can promote better choices among co-workers or community members too—could you talk your boss into installing electric vehicle charging? Or promoting working from home?

- **Rethink Flying:** As we've already discussed, mile-for-mile and dollar-for-dollar, there can be few more effective ways of frying the climate than stepping onto a plane. That said, going entirely flight-free may mean compromising on career goals, or not seeing family. If you are able to avoid flying, it really is a terrific way to reduce your carbon footprint dramatically and, let's face it, flying just sucks anyway. But if you can't, you may still be able to bundle flights, change workplace policies so flying is not the default option, or—if you happen to be wealthy enough to have indulged—switch from business class to economy. Just because you can't go cold turkey just yet doesn't mean you can't get started. Even incremental change—when scaled up across millions of individuals—can lead to disruptive, nonlinear change that will make lower-carbon travel much easier for all of us.

Eat Smarter

Besides travel, diet is an obvious place to reduce your personal climate impacts. And the usual place to start is to think about your meat consumption. That doesn't mean you have to go vegan overnight, however.

- **Become a Reducetarian:** As we saw earlier in the book, there's a growing movement of people who are rethinking their reliance on animal products. Some are hardcore vegans, others vegetarian, and yet others are adopting some form of lower-meat, more plant-centric diet. Find an approach that works for you, because perfection should not be an impediment to action. (In terms of cutting diet-related climate emissions, cutting out beef and other red meats is the most important step.) And once you've found a diet that works for you, look for ways you can start encouraging more plant-based options across your community. (Check out Reducetarian.org for ideas.)

- **Understand Food Waste:** While there's much to love about eating seasonal, local food, the popularity of the farm-to-fork movement has overshadowed another, less sexy side of lower-carbon eating. According to Project Drawdown, cutting food waste—at the household level, in the grocery store, and in the farm fields—is one of the single most important actions society could take to solve the climate crisis. Start paying attention to your food waste at home and, more importantly, apply pressure on grocery store chains, institutions, and policy makers to implement systems-level solutions that keep food from being trashed. (Consider supporting food waste charities like Foodprint.org or FeedbackGlobal.org.) In the UK, at least, you can also now buy beer that's made from waste bread. (ToastAle .com)

- **Don't Forget to Vote:** I might have said this once or twice, but we're not going to solve this as consumers. It is imperative that we engage on the civic level if we are going to transform our food system, or any other aspect of our culture for that matter.

So educate yourself on the politics of food where you live, and then apply pressure to make sure that everyone has access to safe, affordable, and sustainable food that doesn't compromise our climate.

Good Energy

It goes without saying that energy use is a major component of solving the climate crisis. Yet that doesn't mean we all have to (or even can) install solar on our homes. Instead, we need to think about energy use across our entire system—and then find ways to shift it:

- **Go Renewable:** There are many ways to enjoy renewable energy, and not all of them require us to spend a ton of money or own our own home. Many energy providers now offer renewable energy tariffs, meaning you can pay for clean energy that's produced elsewhere. Even if that's not possible where you live, you may be able to donate to non-profits that build renewables. And be sure to bring your enthusiasm to your workplace too— a growing number of businesses, large and small, are realizing that they can save significant amounts of money by installing renewables.
- **Demand Change:** As outlined in the book, there can be few more thorny examples of how vested interests have sabotaged the climate fight than the nefarious double dealings of Big Energy. So regardless of whether or not you are able to go green at home, be sure to find ways to hit them where it hurts. Many of the organizations listed under "Get Organized" above are at the forefront of ensuring accountability for fossil fuel corporations, so be sure to check out their work.
- **Seek Community:** An intriguing aspect of the low-carbon transition that is currently underway is how communities are able to take control of their energy. Whether it's community solar gardens, or neighbors getting together to insulate each others' houses, examples are beginning to take shape all around the

world of literal people power. Be sure to keep an eye out for what's happening in your community. The international network of Transition Communities (TransitionNetwork.org) may be a good place to start.

Money Matters

How we manage and where we keep our money has a surprisingly large impact on the health of the world around us. Whether it's our personal savings or our pension funds, our money has a secret life that can either perpetuate climate destruction, or help fuel solutions instead. Here are some ways you can make sure your money is working for everyone.

- **Divest from Fossil Fuels:** Many pension plans now offer some form of ethical investing options, with some being explicitly fossil-fuel free. It's worth checking the fine print though. The FTSE4Good index, which is my own workplace's ethical plan—offered by Vanguard—does not currently exclude all fossil fuels, as I have a habit of letting them know via social media.

- **Invest in Alternatives:** As the world moves away from fossil fuels, there's a lot of money pouring into clean energy, energy efficiency, vehicle electrification, and other solutions. Whether by picking out individual clean energy stocks, or investing through platforms or companies such as Abundance Investments (UK-based, AbundanceInvestment.com), Triodos Bank (Europe-based, Triodos.com), or Clean Energy Federal Credit Union (US-based, CleanEnergyCU.org), you can make sure your money is actively fueling the transition.

- **Push for More:** Sometimes, it's not possible—or practical—to move all your money, especially if you're participating in a workplace retirement or savings plan. Increasingly, however, there are groups and movements organizing to put pressure on banks, pension funds, and other institutional investors to move their money out of fossil fuels. So be sure to use your position as a customer to make your voice heard.

As I mentioned above, the ideas here are just outlines, suggestions, and starting points. The climate crisis may be all encompassing, complex, and sometimes terrifying. But the silver lining of facing such a multifaceted beast is the fact that we are not short of entry points for starting to make a difference.

None of us can do everything. Few of us will even do all we can. But all of us can and should start doing something. And once we get started, the most important thing is to keep on going. The moment we are in demands nothing less.

Notes

Chapter 1: We Are All Climate Hypocrites Now

1. Shane Gunster et al., "Why Don't You Act Like You Believe It?: Competing Visions of Climate Hypocrisy," *Frontiers in Communication*, November 06, 2018.
2. Jake Tapper, "Al Gore's 'Inconvenient Truth'?—A $30,000 Utility Bill," *ABC News*, February 27, 2007.
3. Juan Cagio Villar et al., "Carbonfeel Project: Calculation, Verification, Certification and Labeling of the Carbon Footprint," Low Carbon Economy, June 27, 2014. www.scirp.org/(S(351jmbntvnsjt1aadkposzje)) /reference/ReferencesPapers.aspx?ReferenceID=1223505
4. Mark Kaufman, "The Carbon Footprint Sham," *Mashable*, July 13, 2020.
5. George Monbiot, "Flying Is Dying," *AlterNet*, March 1, 2006.
6. George Monbiot, "On the Flight Path to Global Meltdown," *The Guardian*, September 21, 2006.
7. "Consumer Action (Boycotts of Sugar and Rum)," The Abolition Project, 2009.
8. Mike Berners-Lee, *There is No Planet B: A Handbook for the Make or Break Years*, (Cambridge University Press, 2019), 6.

Chapter 2: Wants and Needs

1. "Understanding the Irrational Consumer: An Interview with Dan Ariely," *Think With Google*, October 2013.
2. Dan Ariely et al. *Hacking Human Nature for Good*, (Irrational Labs, 2014), 86.
3. Micah Bales, "Let's Go Beyond Consumerism, Citizenship, and Individualism," *Sojourners*, November 5, 2013.
4. Naomi Klein, *On Fire: The (Burning) Case for a Green New Deal*, (Simon & Schuster, 2019), 133.

Chapter 3: How "Green" Lost Its Groove

1. Mary Annaïse Heglar, "I Work in the Environmental Movement. I Don't Care if You Recycle," *Vox*, June 4, 2019.
2. Kate Yoder, "Love it or hate it, Earth Day's just not what it used to be. What happened?" *Grist*, April 22, 2019.
3. Byron Tau, "Pelosi Fires Back at Gingrich over Climate Ad," *Politico*, August 1, 2012.

4. Al Gore, *An Inconvenient Truth: The Planetary Emergency of Global Warming and What We Can Do About It*, (Rodale, 2006), 305.

5. Penelope Green, "The Year Without Toilet Paper," *The New York Times*, March 22, 2007.

6. Colin Beavan, *No Impact Man: The Adventures of a Guilty Liberal Who Attempts to Save the Planet, and the Discoveries He Makes About Himself and Our Way of Life in the Process*, (Picador, 2010), 164.

7. *No Impact Man*, 181.

8. Mark Boyle, *The Moneyless Man: A Year of Freeconomic Living*, (Oneworld Publications, 2010).

9. *No Impact Man*, 167.

10. Lloyd Alter, "Could You Live the 1.5 Degree Lifestyle?" *Treehugger*, January 2, 2020.

Chapter 4: Enough Already

1. Jamie Margolin, "Let's Return Earth Day to its Revolutionary Past," *Sierra Magazine*, March/April 2020, 31.

2. Eric Holthaus, *The Future Earth: A Radical Vision for What's Possible in the Age of Warming*, (HarperOne, 2020), 83.

3. Heather Landy, "A 16-year-old Tells Davos Delegates That if the Planet Dies, She's Blaming Them," *Quartz*, January 25, 2019.

4. "International Rebellion Update #2," Extinction Rebellion website: rebellion.global

5. Sarah Kaplan, "What's the Greenest Way to Travel?" *The Washington Post*, December 12, 2019.

Chapter 5: Guilt Trip

1. katharinehayhoe.com

2. Matt Galloway, "Shaming People Into Fighting Climate Change Won't Work, Says Scientist," *The Current*, CBC radio broadcast, August 19, 2019.

3. "Yum! Brands Announces 'Greener' Paper Policy," MongaBay, April 8, 2013.

4. "Dogwood Alliance and International Paper Find Common Ground," (International Paper, April 10, 2014) [Press Release].

5. Jennifer Jacquet, *Is Shame Necessary?: New Uses for an Old Tool*, (Vintage, 2016) 24.

6. *Is Shame Necessary*, 113.

7. Tim Gore, *Confronting Carbon Inequality*, (Oxfam, 2020).

8. Graham Readfern, "Greyhound Warns Staff They Could be 'In the Crossfire' Over Adani Contract," *The Guardian*, January 22, 2020.

9. Graham Readfern, "Great Barrier Reef Group Severs Ties with Greyhound Over Adani Contract," *The Guardian*, January 22, 2020.

10. Graham Readfern, "Greyhound Cuts Ties with Adani Mine After Backlash from Climate Activists," *The Guardian*, January 28, 2020.

11. Steve Westlake, "Fighting Climate Change: It's Time to Acknowledge

that Even the Smallest Act Can Make a Difference," *The Independent*,
April 24, 2019.

Chapter 6: Big Oil Wants to Talk About Your Carbon Footprint

1. Ben van Beurden, "It is Time to Turn to Nature," *LinkedIn*, April 8, 2019.
2. Josh Gabbatiss. "Shell Predicted Dangers of Climate Change in 1980s and Knew Fossil Fuel Industry was Responsible," *The Independent*, April 09, 2018.
3. Mariah Wojdacz, "Tobacco Companies Pay Big Bucks for Anti-Smoking Campaigns," *LegalZoom*, updated December 8, 2020.
4. Lloyd Alter, "Recycling Is BS; Make Nov. 15 Zero Waste Day, Not America Recycles Day," *Treehugger*, October 11, 2018.
5. Jessica Shankleman, "Mark Carney Lends Weight to Carbon Bubble Theory," *BusinessGreen*, October 13, 2014.
6. Wen Stephenson, *What We're Fighting for Now is Each Other: Dispatches from the Frontlines of Climate Justice*, (Beacon Press, 2018), Loc 94 [Kindle version].
7. Margaryta Kirakosian, "BlackRock Voted Against Climate Resolutions Over 80% of the Time in 2020," *City Wire Selector*, October 1, 2020.
8. Larry Fink, "A Fundamental Reshaping of Finance," *BlackRock.com*, January 15, 2020.
9. twitter.com/balberlin/status/1217186436857892869
10. Sami Grover, "Spain Closes Coal Mines. Mining Unions Celebrate," *Treehugger*, October 30, 2018.
11. "More US Coal-Fired Plants Are Decommissioning as Retirements Continue," *Today in Energy* (U.S. Energy Information Administration), July 26, 2019.
12. Timothy Puko and Rebecca Elliot, "Trump Administration Weighs Aid for Oil Companies," *The Wall Street Journal*, April 21, 2020.

Chapter 7: Corporate "Citizenship" Reimagined

1. patagonia.com/stories/dont-buy-this-jacket-black-friday-and-the-new -york-times/story-18615.html
2. patagonia.com/stories/what-we-do-for-a-living-an-excerpt-from-the -responsible-company/story-18397.html
3. frederickalexander.net/the-shareholder-commons/
4. Umair Irfan, "Exxon is Lobbying for a Carbon Tax. There is, Obviously, a Catch." *Vox*, October 18, 2018.

Chapter 8: Swimming Upstream

1. Kalmus, *Being The Change*, (New Society Publishers, 2017) 260.
2. "The Military Cost of Defending Global Oil Supplies," (Securing America's Future Energy, 2018).
3. Eric Jaffe, "The Real Reason U.S. Gas is So Cheap is Americans Don't Pay the True Cost of Driving," *Bloomberg City Lab*, January 5, 2015

4. David Coady et al. "Global Fossil Fuel Subsidies Remain Large: An Update Based on Country-Level Estimates," (International Monetary Fund, 2019).

5. "Emissions Impossible: How Big Meat and Dairy Are Heating Up the Planet," (GRAIN and Institute for Agriculture and Trade Policy, 2018).

Chapter 9: Focus, Goddammit

1. "South Korea Ferry Disaster: Distrust, Anger Over President's Actions," *The Straits Times*, May 2, 2014.

2. "Sinking of MC Sewol," en.wikipedia.org/wiki/Sinking_of_MV_Sewol

3. S. Mo Jang and Yong Jin Park, "Redirecting the Focus of the Agenda: Testing the Zero-Sum Dynamics of Media Attention in News and User-Generated Media," *International Journal of Communication*, 2017.

4. Cara C. MacInnis and Gordon Hodson, "It Ain't Easy Eating Greens: Evidence of Bias Toward Vegetarians and Vegans from Both Source and Target," *Group Processes & Intergroup Relations*, December 6, 2015.

5. Zaria Gorvett, "The Hidden Biases that Drive Anti-Vegan Hatred," *BBC Future*, February 4, 2020.

6. Maura Judkis, "You Might Think There Are More Vegetarians Than Ever. You'd Be Wrong." *The Washington Post*, August 3, 2018.

7. "The Power of Meat 2019," FMI and Foundation for Meat and Poultry Education and Research, 2019.

8. Elaine Watson, "Impossible Foods: 'Our goal is to produce a full range of meats and dairy products for every cultural region in the world,'" *Food Navigator*, April 12, 2018.

9. Umair Irfan, "Air Travel is a Huge Contributor to Climate Change. A New Global Movement Wants You to be Ashamed to Fly." *Vox*, November 30, 2019.

10. ec.europa.eu/clima/policies/transport/aviation_en

11. Stefan Gösling and Andreas Humpe, "The Global Scale, Growth and Distribution of Aviation: Implications for Climate Change," *Global Environmental Change*, November 2020.

12. "Sweden Sees Rare Fall in Air Passengers, as Flight-Shaming Takes Off," *BBC*, January 10, 2020.

13. Nicole Lyn Pesce, "How Greta Thunberg and 'Flygskam' are Shaking the Global Airline Industry," *Market Watch*, December 19, 2019.

14. Sami Grover, "Huge Shift from Planes to Trains on London-Scotland Routes," *Treehugger*, October 11, 2018.

15. hiltner.english.ucsb.edu/index.php/ncnc-guide/#intro

16. "Business Travel by the Numbers" (Trondent Development Group)

Chapter 10: What Difference Does It Make?

1. facebook.com/watch/?v=491468944888136

2. Pete Jordan, *In the City of Bikes: The Story of the Amsterdam Cyclist*, (Harper Perennial, 2013), 283.

3. Ibid. 345.

4. Ibid, 350.

5. Gloria Kurnik, "How Amsterdam Became a Heaven for Cyclists," *Then This Happened*, Bloomberg (Online Video), October 25, 2018.

6. Michael E. Mann, "Lifestyle Changes Aren't Enough to Save the Planet. Here's What Could," *Time*, September 12, 2019.

7. Ajah Hales, "Instead of White Guilt, We Need Racial Resilience," *The Salve*, August 19, 2019.

8. Jocelyn Timperley, "Who is Really to Blame for Climate Change," *BBC Future*, June 18, 2020.

Chapter 11: Climate Hypocrites Unite!

1. Simon Evans, "Analysis: UK's CO_2 Emissions Have Fallen 29 Per Cent Over the Past Decade," Carbon Brief, March 3, 2020.

2. Leah Stokes, *Short Circuiting Policy: Interest Groups and the Battle Over Clean Energy and Climate Policy in the American States*, (Oxford University Press, 2020), 18.

3. Jussi Impiö et al. "Pathways to 1.5 Degree Lifestyles by 2030," *Sitra*, September 4, 2020.

4. Klein, *On Fire*, 135.

5. twitter.com/aldatweets/status/1301552739588214786

Index

About the Author

SAMI GROVER is a green lifestyle blogger and self-confessed eco-hypocrite. He has spent most of his life trying to live a greener lifestyle and has written more than 2,000 articles covering everything from electric bike ownership to peeing on your compost heap. Yet he has only been marginally successful in reducing his own environmental impact. Active in the sphere of good-for-the-world business, he has developed branding projects for clients including Burt's Bees, Dogwood Alliance, and Jada Pinkett Smith. He believes that, in order to make a difference, each of us has to identify our greatest point of leverage and focus our efforts there. He lives in Durham, North Carolina, with his wife and kids.